普通高等教育电气工程自动化系列教材

电气工程与自动化专业英语

第2版

主　编　王　伟　张殿海　裴素萍
副主编　马景兰　韩肖清

机械工业出版社

本书分为电气工程基础、控制理论与技术、电机与电器、电力系统 4 大部分，共 20 章，涵盖了电气工程与自动化专业的主要专业基础理论内容，其中包括目前电气工程领域研究的热点内容。本书专业性和实用性强，注重从实际应用出发培养学生的专业英语阅读、表述和写作能力。每章后均补充了必要的专业英语词汇、短语及句子注释，便于不同读者对各章知识点的正确理解和自学。

本书既可作为高等院校电气工程与自动化专业学生的专业英语教材，也可作为相关专业学生和工程技术人员的参考用书。

图书在版编目（CIP）数据

电气工程与自动化专业英语/王伟，张殿海主编. —2 版. —北京：机械工业出版社，2017.12（2023.6 重印）
普通高等教育电气工程自动化系列教材
ISBN 978-7-111-58312-7

Ⅰ. ①电… Ⅱ. ①王… ②张… Ⅲ. ①电工技术—英语—高等学校—教材②自动化技术—英语—高等学校—教材 Ⅳ. ①TM②TP2

中国版本图书馆 CIP 数据核字（2017）第 253793 号

机械工业出版社（北京市百万庄大街 22 号 邮政编码 100037）
策划编辑：王雅新 责任编辑：王雅新 杨 洋 王小东
责任校对：尹 君 封面设计：张 静
责任印制：刘 媛
涿州市般润文化传播有限公司印刷
2023 年 6 月第 2 版第 8 次印刷
184mm×260mm · 15 印张 · 365 千字
标准书号：ISBN 978-7-111-58312-7
定价：36.00 元

电话服务　　　　　　　　　网络服务
客服电话：010-88361066　　机　工　官　网：www.cmpbook.com
　　　　　010-88379833　　机　工　官　博：weibo.com/cmp1952
　　　　　010-68326294　　金　书　网：www.golden-book.com
封底无防伪标均为盗版　　　机工教育服务网：www.cmpedu.com

前　言

本书第1版自2010年8月出版以来，被多所高校的电气工程及其自动化专业及其他电气类专业作为教材用于专业英语的教学中。编者在教学过程中也收到了使用本书的教师们的反馈信息，他们在对本书给予充分肯定的同时，也提出了宝贵的建议和修改意见，在此表示由衷的感谢。

本书第2版在保持本书专业知识点突出、实用性强的基础上，主要做了以下三方面的修订和补充：

1) 补充了与电气工程及其自动化专业发展相关的内容，如特种电机、微电网技术等方面的内容。

2) 对部分章节的内容安排进行了调整，使整本书各知识点内容在突出专业特点和重点的同时，内容衔接更紧密。

3) 为了便于相关专业和不同年级读者的使用，本书每章后增加了对重点或难点句子的中文注释，可帮助读者理解和学习有关知识点内容。

本书第2版由北京石油化工学院、沈阳工业大学、太原理工大学、中原工学院几所高校多年从事专业英语教学工作的老师联合修订。第2版共包括电气工程基础（第1~5章）、控制理论与技术（第6~10章）、电机与电器（第11~15章）及电力系统（第16~20章）4个部分，整体结构与本书第1版保持一致。参与本书修订工作的有王伟、韩肖清、张殿海、马景兰和裴素萍5位老师。全书由王伟和马景兰老师统稿。张艳丽老师由于工作原因未参加本版的修订工作，相关内容由沈阳工业大学张殿海老师负责修订，在此也对张艳丽老师在本书第1版中所做的工作表示感谢。此外，北京石油化工学院的李伟、太原理工大学的王鹏敏和河南省建筑科学研究院的刘禹等也对本书的修订做了许多工作，在此一并表示感谢。

由于编者水平和经验有限，书中难免存在疏漏和不足之处，敬请读者批评指正。

编　者

目 录

前言

PART 1　FUNDAMENTALS OF ELECTRIC ENGINEERING ……1

Chapter 1　Circuit Fundamentals …… 1
1.1　Electrostatic Charges …… 1
1.2　Conductors, Insulators and Semiconductors …… 2
1.3　Current, Voltage and Resistance …… 2
1.4　Measuring Resistance, Voltage and Current …… 5
1.5　DC Series Electrical Circuit …… 7
1.6　Alternating Current Voltage …… 8

Chapter 2　Analog Electronics …… 14
2.1　Introduction …… 14
2.2　Operational Amplifiers …… 14
2.3　Differential Amplifiers …… 17
2.4　Integrator and Differentiator …… 19
2.5　Active Filters …… 20

Chapter 3　Digital Electronics …… 26
3.1　Introduction …… 26
3.2　Digital Number Systems …… 26
3.3　Binary Logic Circuits …… 29
3.4　Combination Logic Gates …… 30
3.5　Timing and Storage Elements …… 32

Chapter 4　Power Electronics Technology …… 39
4.1　Introduction …… 39
4.2　Applications and the Roles of Power Electronics …… 39
4.3　Structure of Power Electronics Interface …… 43
4.4　Voltage-Link Structure Applications …… 45
4.5　Recent and Potential Advancements …… 47

Chapter 5　Magnetism and Electromagnetism …… 52
5.1　Introduction …… 52
5.2　Permanent Magnets …… 52
5.3　Magnetic Field around Conductors and a Coil …… 53
5.4　Ohm's Law for Magnetic Circuits …… 56
5.5　Domain Theory of Magnetism …… 56
5.6　Electricity Produced by Magnetism …… 57

PART 2 CONTROL THEORY AND TECHNOLOGY ... 62

Chapter 6 Knowledge of Control Theory ... 62
6.1　What Is Control ... 62
6.2　Feedback ... 63
6.3　PID Control ... 65
6.4　Adaptive Control ... 66

Chapter 7 Motor Drives and Controls ... 73
7.1　DC Motor Drives ... 73
7.2　Inverter-Fed Induction Motor Drives ... 79

Chapter 8 Programmable Logic Controller Technology ... 90
8.1　Introduction ... 90
8.2　PLC Operation Process ... 93
8.3　PLC Maintenance Management ... 94
8.4　The Application Future of PLC ... 96

Chapter 9 Single Chip Microcomputer Control Technology ... 102
9.1　Foundation ... 102
9.2　A Single-Chip Microcomputer Integrated Circuit ... 105
9.3　Digital Signal Processors ... 107

Chapter 10 Computer Networking Basics ... 112
10.1　Foundation ... 112
10.2　Applications ... 112
10.3　Requirements ... 113
10.4　Links, Nodes and Clouds ... 114
10.5　Network Architecture ... 116
10.6　Network Security ... 120

PART 3 ELECRICAL MACHINES AND DEVICES ... 125

Chapter 11 Direct-Current Machines ... 125
11.1　Introduction ... 125
11.2　Basic Structural Feature ... 127
11.3　Effect of Armature MMF ... 132

Chapter 12 Three-Phase Induction Motors ... 137
12.1　Introduction ... 137
12.2　Construction of Three-Phase Induction Motors ... 138
12.3　Principle of Operation ... 140
12.4　Equivalent Circuit ... 142

Chapter 13 Synchronous Machines ... 148
13.1　Introduction ... 148
13.2　Principle of Operation ... 149

Chapter 14 Transformers ... 156
14.1　Introduction ... 156

14.2　Transformer Construction ……………………………………………………………… 157
14.3　Ideal Transformers ……………………………………………………………………… 161

Chapter 15　Permanent Magnet Machines …………………………………………… 165

15.1　Introduction ……………………………………………………………………………… 165
15.2　Introduction of PM Materials …………………………………………………………… 165
15.3　Classification of PM Motors …………………………………………………………… 167
15.4　Operational Principle of PM Motors …………………………………………………… 168

PART 4　POWER SYSTEMS …………………………………………………………… 175

Chapter 16　Operating Characteristics of Modern Power Systems ……………… 175

16.1　Transmission and Distribution Systems ……………………………………………… 175
16.2　Power System Controls ………………………………………………………………… 177
16.3　Generator-Voltage Control ……………………………………………………………… 178
16.4　Turbine-Governor Control ……………………………………………………………… 180
16.5　Load-Frequency Control ………………………………………………………………… 182
16.6　Optimal Power Flow …………………………………………………………………… 183
16.7　Power System Stability ………………………………………………………………… 184

Chapter 17　Generating Plants ………………………………………………………… 188

17.1　Electric Energy ………………………………………………………………………… 188
17.2　Fossil-Fuel Plants ……………………………………………………………………… 190
17.3　Nuclear Power Plants …………………………………………………………………… 192
17.4　Hydroelectric Power Plants …………………………………………………………… 193
17.5　Wind Power Systems …………………………………………………………………… 194
17.6　Photovoltaic Systems …………………………………………………………………… 197

Chapter 18　Concepts and Models for Microgeneration and Microgrids ………… 201

18.1　Introduction ……………………………………………………………………………… 201
18.2　The Foundation of Microgrid Concept ………………………………………………… 203
18.3　The Microgrid Operational and Control Architecture ………………………………… 206

Chapter 19　High Voltage Insulation …………………………………………………… 213

19.1　Introduction ……………………………………………………………………………… 213
19.2　Lightning ………………………………………………………………………………… 213
19.3　Switching Surges ………………………………………………………………………… 215
19.4　Insulation Coordination ………………………………………………………………… 216

Chapter 20　System Protection ………………………………………………………… 222

20.1　Introduction ……………………………………………………………………………… 222
20.2　Protection of Radial Systems …………………………………………………………… 223
20.3　System with Two Sources ……………………………………………………………… 224
20.4　Impedance (Distance) Relays …………………………………………………………… 226
20.5　Differential Protection of Generators ………………………………………………… 228
20.6　Differential Protection of Transformers ……………………………………………… 230
20.7　Computer Relaying ……………………………………………………………………… 232

PART 1 FUNDAMENTALS OF ELECTRIC ENGINEERING

Chapter 1 Circuit Fundamentals

1.1 Electrostatic Charges

Protons and electrons are parts of atoms that make up all things in our world. The positive charge of a proton is similar to the negative charge of an electron. However, a positive charge is the opposite of a negative charge. These charges are called electrostatic charges. Each charged particle is surrounded by an electrostatic field. [1]

The effect that electrostatic charges have on each other is very important. They either repel (move away) or attract (come together) each other. It is said that like charges repel and unlike charges attract. [2]

The atoms of some materials can be made to gain or lose electrons. The material then becomes charged. One way to do this is to rub a glass rod with a piece of silk cloth. The silk cloth pulls electrons (-) away from the glass. The glass rod loses electrons, so it now has a positive (+) charge. Since the silk cloth gains new electrons, it now has a negative (-) charge. Another way to charge a material is to rub a rubber rod with fur.

It is also possible to charge other materials when they are brought close to another charged object. If a charged rubber rod is touched against another material, the second material may become charged. [3] Remember that materials are charged due to the movement of electrons and protons.

Charged materials affect each other due to lines of force. These imaginary lines cannot be seen. However, they exert a force in all directions around a charged material. Their force is similar to the force of gravity around the earth. This force is called a gravitational field.

Most people have observed the effect of static electricity. Whenever objects become charged, it is due to static electricity. A common example of static electricity is lightning. Lightning is caused by a difference in charge (+ and -) between the earth's surface and the clouds during a storm. The arc produced by lightning is the movement of charges between the earth and the clouds. Another common effect of static electricity is being "shocked" by a doorknob after walking across a carpeted floor. Static electricity also causes clothes taken from a dryer to cling together and hair to stick to a comb.

Electrical charges are used to filter dust and soot in devices called electrostatic filters. Electrostatic precipitators are used in power plants to filter the exhaust gas that goes into the air.[4] Static electricity is also used in the manufacture of sandpaper and in the spray painting of automobiles. A device called an electroscope is used to detect a negative or positive charge.

1.2 Conductors, Insulators and Semiconductors

1.2.1 Conductors

A material through which current flows is called a conductor. A conductor passes electric current very easily. Copper and aluminum wire are commonly used as conductors. Conductors are said to have low resistance to electrical current flow. Conductors usually have three or fewer electrons in the outer orbit of their atoms.[5] Remember that the electrons of an atom orbit around the nucleus. Many metals are electrical conductors. Each metal has a different ability to conduct electric current. Materials with only one outer orbit or valence electron (gold, silver, copper) are the best conductors. For example, silver is a better conductor than copper, but it is too expensive to use in large amounts. Aluminum does not conduct electrical current as well as copper, but it is cheaper and lighter than other conductors, so it was commonly used in the past. Copper is used more than any other conductor at present.

1.2.2 Insulators

There are some materials that do not allow electric current to flow easily. The electrons of materials that are insulators are difficult to release. In some insulators, their valence shells are filled with eight electrons. The valence shells of others are over half-filled with electrons. The atoms of materials that are insulators are said to be stable.[6] Insulators have high resistance to the electric currents that pass through them. Some examples of insulators are plastic and rubber.

1.2.3 Semiconductors

Materials called semiconductors have become very important in electronics. Semiconductor materials are neither conductors nor insulators. Their classification also depends on the number of electrons that their atoms have in their valence shells. Semiconductors have 4 electrons in their valence shells.[7] Remember that conductors have outer orbits less than half-filled and insulators ordinarily have outer orbits more than half-filled. Some common types of semiconductor materials are silicon, germanium, and selenium.

1.3 Current, Voltage and Resistance

We depend on electricity to do many things that are sometimes taken for granted. There are three basic electrical terms which must be understood, current, voltage, and resistance.

1.3.1 Current

Static electricity is caused by stationary charges. However, electrical current is the motion of electrical charges from one point to another. Electric current is produced when electrons are removed from their atoms.[8] A force or pressure applied to a material causes electrons to be removed. The movement of electrons from one atom to another is called electric currentflow.

1. Current Flow

The usefulness of electricity is due to its electric current flow. Current flow is the movement of electrical charges along a conductor. Static electricity, or electricity at rest, has some practical uses due to electrical charges. Electric current flow allows us to use electrical energy to do many types of work.

The movement of valence shell electrons of conductors produces electrical current. The outer electrons of the atoms of a conductor are called free electrons. Energy released by these electrons as they move allows work to be done.[9] As more electrons move along a conductor, more energy is released. This is called an increased electric current flow.

To understand how current flow takes place, it is necessary to know about the atoms of conductors. Conductors, such as copper, have atoms that are loosely held together. Copper is said to have atoms connected together by metallic bonding. A copper atom has one valence shell electron, which is loosely held to the atom. These atoms are so close together that their valence shells overlap each other. Electrons can easily move from one atom to another. In any conductor the outer electrons continually move in a random manner from atom to atom. The random movement of electrons does not result in current flow, since electrons must move in the same direction to cause current flow.

When electric charges are placed on each end of a conductor, there is a difference in the charges at each end of the conductor. Current flow takes place because the free electrons move in one direction. Remember that like charges repel and unlike charges attract. Free electrons have a negative charge, so they are repelled by the negative charges at the same side. The free electrons are attracted to the positive charges on the other side and move to the other side from one atom to another. If the charges on each end of the conductor are increased, more free electrons will move. This increased movement causes more electric current flow.

Current flow is the result of electrical energy caused as electrons change orbits. This impulse moves from one electron to another. When one electron moves out of its valence shell, it enters another atom's valence shell. An electron is then repelled from that atom. This action goes on in all parts of a conductor. Remember that electric current flow produces a transfer of energy.

2. Electronic Circuits

Current flow takes place in electronic circuits. A circuit is a path or conductor for electric current flow. Electric current flows only when it has a complete, or closed-circuit, path. There must be a source of electrical energy to cause current to flow along a closed path.[10] The electrical energy is converted into more useful energy, for example the light energy.

Electric current cannot flow if a circuit is open. An open circuit does not provide a complete path for current flow. Free electrons of the conductor would no longer move from one atom to another. An example of an open circuit is a "burned-out" light bulb. Actually, the filament (the part that produces light) has become open. The open filament of a light bulb stops current flow from the source of electrical energy. This causes the bulb to stop burning, or producing light.

Another common circuit term is a short circuit. A short circuit, which can be very harmful, occurs when a conductor connects directly across the terminals of an electrical energy source. For safety purposes, a short circuit should never happen because short circuits cause too much current to flow from the source. If a wire is placed across a battery, a short circuit occurs. The battery would probably be destroyed and the wire could get hot or possibly melt due to the short circuit.

1.3.2 Voltage

Water pressure is needed to force water along a pipe. Similarly, electrical pressure is needed to force current along a conductor. If a motor is rated at 220 V, it requires 220 V of electrical pressure applied to the motor to force the proper amount of current through it. More pressure would increase the current flow and less pressure would not force enough current to flow. The motor would not operate properly with too much or too little voltage. An electrical energy source such as a battery or generator produces current flow through a circuit. As voltage is increased, the amount of current in the circuit is also increased. Voltage is also called electromotive force (EMF).[11]

1.3.3 Resistance

The opposition to current flow in electrical circuits is called resistance. Resistance is not the same for all materials. The number of free electrons in a material determines the amount of opposition to current flow. Atoms of some materials give up their free electrons easily. These materials offer low opposition to current flow. Other materials hold their outer electrons and offer high opposition to current flow.[12]

Electric current is the movement of free electrons in a material. Electric current needs a source of electrical pressure to cause the movement of free electrons through a material. An electric current will not flow if the source of electrical pressure is removed. A material will not release electrons until enough force is applied. With a constant amount of electrical force (voltage) and more opposition (resistance) to current flow, the number of electrons flowing (current) through the material is smaller. With constant voltage, current flow is increased by decreasing resistance. Decreased current results from more resistance. By increasing or decreasing the amount of resistance in a circuit, the amount of current flow can be changed.

Even very good conductors have some resistance, which limits the flow of electric current through them. The resistance of any material depends on four factors:

1) The material of which it is made;
2) The length of the material;
3) The cross-sectional area of the material;

4) The temperature of the material.

The material of which an object being made affects its resistance. The ease with which different materials give up their outer electrons is very important in determining resistance. Silver is an excellent conductor of electricity. Copper, aluminum, and iron have more resistance but are more commonly used, since they are less expensive. All materials conduct an electric current to some extent, even though some (insulators) have very high resistance.

Length also affects the resistance of a conductor. The longer a conductor, the greater the resistance is. A material resists the flow of electrons because of the way in which each atom holds onto its outer electrons. The more material that is in the path of an electric current, the less current flow the circuit will have. If the length of a conductor is doubled, there is twice as much resistance in the circuit.

Another factor that affects resistance is the cross-sectional area of a material. The greater the cross-sectional area of a material is, the lower the resistance. If two conductors have the same length but twice the cross-sectional area, there is twice as much current flow through the wire with the larger cross-sectional area.

Temperature also affects resistance. For most materials, the higher the temperature, the more resistance it offers to the flow of electric current. This effect is produced because a change in the temperature of a material changes the ease with which a material releases its outer electrons. A few materials, such as carbon, have lower resistance as the temperature increases. The effect of temperature on resistance varies with the type of material. The effect of temperature on resistance is the least important of the factors that affect resistance.

1.4 Measuring Resistance, Voltage and Current

Another important activity in the study of electronics is measurement. Measurements are made in many types of electronic circuits. The proper ways of measuring resistance, voltage, and current should be learned.

Volt-Ohm-Milliammeters (VOMs), or multimeters, are the most used meters for doing electronic work. A VOM is often used to measure resistance, voltage, and current by electronics technicians. The type of measurement is changed by adjusting the "function-select switch" to the desired measurement.

The test leads used with the VOM are ordinarily black and red. These colors are used to help identify which lead is the positive and which is the negative side of the meter. Red indicates positive (+) polarity and black indicates the negative (−) polarity. The red test lead is put in the hole, or jack, marked V − Ω − A, or volts − ohms − amperes. The black test lead is put in the hole, or jack, labeled-COM.

1.4.1 Measuring Resistance

Many important electrical tests may be made by measuring resistance. Resistance is opposition to the flow of current in an electrical circuit. The current that flows in a circuit depends upon the amount of resistance in that circuit. You should learn to measure resistance in an electronic circuit by using a meter.

The ohmmeter ranges of a VOM, or multimeter, is used to measure resistance. The basic unit of resistance is the ohm (Ω). When the test leads are touched together, or "shorted", the meter is operational, indicating zero (0) ohms.

Never measure the resistance of a component until it has been disconnected, if not the reading may be wrong. Voltage should never be applied to a component when its resistance is being measured.

1.4.2 Measuring Voltage

Voltage is applied to electrical equipment to cause it to operate. It is important to be able to measure voltage to check the operation of equipment. Many electrical problems develop when either too high or too low voltage is applied to the equipment. Voltage is measured in volts (V). A voltmeter ranges of a VOM is used to measure voltage in an electrical circuit.

When making voltage measurements, connect the red and black test leads to the meter by putting them into the proper jacks. The red test lead should be put into the jack labeled V-Ω-A. The black test lead should be put into the jack labeled-COM.

The proper voltage range is chosen. Before making any measurements, adjust the function-select switch to the highest voltage range. The value of the range being used is the maximum value of voltage that can be measured on that range. For example, when the range selected is 12 V, the maximum voltage that the meter can measure is 12 V. Any voltage above 12 V could damage the meter. To measure a voltage that is unknown, start by using the highest range on the meter. Then slowly adjust the range downward until a voltage reading is indicated on the right side of the meter scale.

AC (Alternating Current) voltage is measured with a VOM and polarity of the meter leads is not important because AC changes direction. Matching the polarity of the meter to the voltage polarity is very important when measuring DC (Direct Current) voltage, since direction current flows only in one direction. Meter polarity is simple to determine. The positive (+) red test lead is connected to the positive side of the DC voltage being measured. The negative (−) black test lead is connected to the negative side of the DC voltage being measured. The meter is always connected across (in parallel with) the DC voltage or AC voltage being measured.

1.4.3 Measuring Current

Current flows through a complete electrical circuit when voltage is applied. Many important tests are made by measuring current flow in electrical circuits. The current values in an electrical circuit depend on the amount of resistance in the circuit. The basic unit of current is ampere (A). Current is commonly measured in units called milliamperes (mA) and microamperes (μA). Learning to use an ammeter to measure current in an electrical circuit is very important.

The function-select switch of a VOM may be adjusted to any one of the current ranges. The value of the current set on the range is the maximum value that can be measured on that range. For example, when the function-select switch is placed in the 120 mA range, the meter is capable of measuring up to 120 mA of current. The function-select switch should first be adjusted to the highest range of direct current. Current is measured by connecting the meter into a circuit, i. e. connecting the meter in series with the circuit. Current flows from a voltage source when some device that has resistance is connected to the source. [13]

Always remember the following safety tips when measuring current:

1) Turn off the voltage before connecting the meter in order not to get an electrical shock. This is an important habit to develop.

2) Set the meter to its highest current range.

3) Disconnect a wire from the circuit and put the meter in series with the circuit.

4) Use the proper meter polarity.

1.5 DC Series Electrical Circuit

There are three types of electrical circuits. They are called series circuits, parallel circuits, and combination circuits. The easiest type of circuit to understand is the series circuit. [14] Series circuits are different from other types of electrical circuits. In a series circuit, there is only one path for current to flow. Since there is only one current path, the current flow is the same value in any part of the circuit. The voltages in the circuit depend on the resistance of the components in the circuit. When a series circuit is opened, there is no path for current flow. Thus, the circuit does not operate. It is important to remember the characteristics of a series circuit.

In the circuit examples that follow, subscripts (such as in R_T, V_T, I_1) are used to identify electrical components in circuit diagrams. The circuit shown in Fig. 1-1 has two resistors and a battery. The resistors are labeled R_1 and R_2. The subscripts identify each of these resistors. Subscripts also aid in making measurements. The voltage drop across resistor R_1 is called voltage drop V_1. The term total is represented by the subscript T, such as in V_T, which is total voltage applied to a circuit. The current measurement I_2 is the current through resistor R_2. Total current is I_T. The voltage drop across R_2 is called V_2.

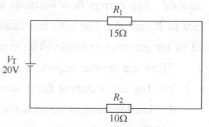

Fig. 1-1 Series Electrical Circuit

Subscripts are also valuable in troubleshooting and repair of electronic equipment. It would be impossible to isolate problems in equipment without components that are easily identified.

The main characteristic of a series circuit is that it has only one path for current flow. In the circuit shown in Fig. 1-1, current flows from the positive side of the voltage source through resistor R_1, through resistor R_2, and then to the negative side of the voltage source. Another characteristic of a series circuit is that the current is the same everywhere in the circuit. A VOM can be placed into the circuit to measure current.

In any series circuit, the sum of the voltage drops is equal to the voltage applied to the circuit. The circuit shown in Fig. 1-1 has voltage drops of 12 V plus 8 V, which is equal to 20 V. Another characteristic of a series circuit is that its total resistance to current flow is equal to the sum of all resistance in the circuit. In the circuit shown in Fig. 1-1, the total resistance of the circuit is the sum of the two resistances if the internal resistance is not considered, so the total resistance is equal to 15 Ω plus 10 Ω, i.e. 25 Ω.

When a series circuit is opened, there is no longer a path for current flow. The circuit will not operate. In the circuit of Fig. 1-1, if R_1 and R_2 are replaced respectively by Lamp 1 and Lamp 2, when the Lamp 1 is burned out, its filament is open. Since a series circuit has only one current path, that path is broken. No current flows in the circuit. The Lamp 2 will not work either. If one light burns out, the others go out also.

Ohm's law is used to explain how a series circuit operates. In the circuit of Fig. 1-1, the total resistance is equal to 15 Ω plus 10 Ω, i.e. 25 Ω. The applied voltage is 20V. Current is equal to voltage divided by resistance, i.e. $I = V/R$. In the circuit shown, current is equal to 20 V divided by 25 Ω, which is 0.8 A. If a current meter were connected into this circuit, the current measurement would be 0.8 A. Voltage drops across each of the resistors may also be found. Voltage is equal to current times resistance ($V = I \times R$). The voltage drop across R_1 (V_1) is equal to the current through R_1 (0.8 A) times the value of R_1 (15 Ω), which is 0.8 A × 15 Ω, i.e. 12 V. The voltage drop across R_2 (V_2) equals 0.8 A × 10 Ω, i.e. 8 V. The sum of these voltage drops is equal to the applied voltage. To check these values, add 12V plus 8V, which is equal to 20 V.

If another resistance is added to a series circuit, the resistance of the total circuit will be increased. The current flow becomes smaller. The circuit now has R_3 (a 5 Ω resistor) added in series to R_1 and R_2. The total resistance is now 15 Ω + 10 Ω + 5 Ω, i.e. 30 Ω, compared to 25 Ω in the previous example. The current is now 0.67 A.

There are several important characteristics of series circuits:

1) The same current flows through each part of a series circuit.
2) The total resistance of a series circuit is equal to the sum of the individual resistances.
3) The voltage applied to a series circuit is equal to the sum of the individual voltage drops.
4) The voltage drop across a resistor in a series circuit is directly proportional to the size of the resistor.
5) If the circuit is broken at any point, no current will flow.

1.6 Alternating Current Voltage

When an Alternating Current (AC) source is connected to some type of load, current direction changes several times in a given unit of time. Remember that DC flows in one direction only. This waveform is called an AC sine wave. When the AC generator shaft rotates one complete revolution, or 360°, one AC sine wave is produced. Note that the sine wave has a positive peak at 90° and then decreases to zero at 180°. It then increases to a peak negative voltage at 270° and then decreases to zero at 360°. The cycle then repeats itself. Current flows in one direction during the positive part and in the opposite direction during the negative half-cycle.

1.6.1 Parameters Associated with AC Sine Wave

If the time required for an AC generator to produce five cycles were 1 s, the frequency of the AC would be 5 cycles per second. AC generators at power plants in the United States operate at a frequency of 60 cycles per second, or 60 hertz (Hz). The hertz is the international unit for frequency measurement. If 60 AC sine waves are produced every second, a speed of 60 revolutions per second is needed. This produces a frequency of 60 cycles per second.

Fig. 1-2 shows several voltage values associated with single-phase AC. Among these are peak positive, peak negative, and peak-to-peak AC values. Peak positive is the maximum positive voltage reached during a cycle of AC. Peak negative is the maximum negative voltage reached. Peak-to-peak is the voltage value from peak positive to peak negative. These values are important to know when working with radio and television amplifier circuits. For example, the most important AC value is called the effective, or measured, value. This value is less than the peak positive value. A common AC voltage is 220 V, which is used in homes. This is an effective value voltage. Its peak value is about 311V. [15] Effective value of AC is defined as the AC voltage that will do the same amount of work as a DC voltage of the same value. For instance, a lamp should produce the same amount of brightness with a 10V AC effective value as with 10 V DC applied. When AC voltage is measured with a meter, the reading indicated is effective value.

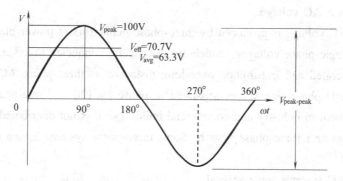

Fig. 1-2 AC Voltage Waveform for a Cycle

In some cases, it is important to convert oneAC value to another. For instance, the voltage rating of electronic devices must be greater than the peak AC voltage applied to them. If 220 V AC is the measured voltage applied to a device, the peak voltage is about 311 V, so the device must be rated over 311 V rather than 220 V.

To determine peak AC, when the measured or effective value is known, the formula

$$V_{peak} = 1.414 V_{eff}$$

is used. When220 V is multiplied by the conversion factor 1.414, the peak voltage is found to be about 311 V.

Two other important terms are RMS value and average value. RMS stands for *root mean square* and is equal to 0.707 times peak value. RMS refers to the mathematical method used to determine effective voltage. RMS voltage and effective voltage are the same. Average voltage is the mathematical average of all instantaneous voltages that occur at each period of time throughout an alternation. The average value is equal to 0.636 times the peak value.

The term *phase* refers to time, or the difference between one point and another. If two sine-wave voltages reach their zero and maximum values at the same time, they are in phase.[16] Fig. 1-3 shows two AC voltages, 1 and 2 that are in phase. If two voltages reach their zero and maximum values at different times, they are out of phase. Fig. 1-3 shows two AC voltages, 1 and 3 that are out of phase. Phase difference is given in degrees. The voltages shown are out of phase by 90°.

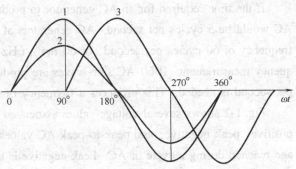

Fig. 1-3　The Phase Waveforms of AC Voltages

1.6.2　Single-Phase and Three-Phase AC Voltages

Single-phase AC voltage is produced by single-phase AC generators or is obtained across two power lines of a three-phase system. A single-phase AC source has a hot wire and a neutral wire. Single-phase AC voltage is used for low-power applications. The type of power distributed to our homes is single-phase AC voltage.

Three-phase AC voltage is produced by three-phase generators at power plants and is a combination of three single-phase voltages, which are electrically connected together. Almost all electrical power is generated and transmitted over long distances as three-phase AC. This voltage is similar to three single-phase sine waves separated in phase by 120°. Three-phase AC is used to power large equipment in industry and commercial buildings. It is not distributed to homes. There are three power lines on a three-phase system. Some three-phase systems have a neutral connection and others do not.[17]

Three-phase AC systems have several advantages over single-phase systems. In a single-phase system, the power is said to be pulsating. The peak values along a single-phase AC sine wave are separated by 360°, for example, the U phase as shown in Fig. 1-4. This is similar to a one-cylinder gas engine. A three-phase system is some what like a multi-cylinder gas en-

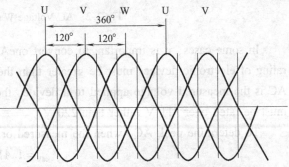

Fig. 1-4　Waveform for Three-Phase AC Voltages

gine. The power is steadier since one cylinder is compressing while the others are not. The power of one separate phase is pulsating, but the total power is more constant. The peak values of three-phase AC are separated by 120°, as shown in Fig. 1-4. These make three-phase AC power more desirable to use.

The power ratings of motors and generators are greater when three-phase AC powers are used. For a certain frame size, the rating of a three-phase AC motor is almost 50% larger than a similar single-phase AC motor.[18]

New Words and Expressions

1. electrostatic   *adj.* 静电的
2. charge   *n.* 电荷
3. positive   *adj.* 正的
4. negative   *adj.* 负的
5. electron   *n.* 电子
6. proton   *n.* 质子
7. gravitational   *adj.* 引力的
8. precipitator   *n.* 电滤器
9. power plant   发电厂
10. current   *n.* 电流
11. conductor   *n.* 导体
12. insulator   *n.* 绝缘体
13. semiconductor   *n.* 半导体
14. valence shell   价电子层
15. silicon   *n.* 硅
16. germanium   *n.* 锗
17. selenium   *n.* 硒
18. voltage   *n.* 电压
19. resistance   *n.* 电阻
20. circuit   *n.* 电路
21. electromotive force (EMF)   电动势
22. repel   *v.* 排斥
23. multimeter   *n.* 万用表
24. milliammeter   *n.* 毫安表
25. ammeter   *n.* 安培表
26. ohmmeter   *n.* 欧姆表
27. voltmeter   *n.* 电压表
28. polarity   *n.* 极性
29. parallel   *n.* 并联
30. series   *n.* 串联
31. filament   *n.* 灯丝
32. resistor   *n.* 电阻器
33. phase   *n.* 相位
34. hot wire   火线
35. neutral wire   中线
36. electrical power   电源

Notes

[1] However, a positive charge is the opposite of a negative charge. These charges are called electrostatic charges. Each charged particle is surrounded by an *electrostatic field*.

然而，正电荷与负电荷是相反的。这些电荷被称为静电荷。每一个带电粒子都被静电场围绕着。

[2] The effect that electrostatic charges have on each other is very important. They either repel (move away) or attract (come together) each other. It is said that like charges repel and unlike charges attract.

静电荷彼此之间的影响是很重要的。它们或者排斥（远离），或者吸引（聚集）。这就是通常人们所说的同性相斥，异性相吸。

[3] It is also possible to charge other materials when they are brought close to another charged object. If a charged rubber rod is touched against another material, the second material may become charged.

当一些材料与另一带电体接近时，就会带上电荷。如果带电的橡胶棒与另一材料相接触，就可以使该材料带上电荷。

[4] Electrical charges are used to filter dust and soot in devices called electrostatic filters. Electrostatic precipitators are used in power plants to filter the exhaust gas that goes into the air.

在被称为静电过滤器的设备中，使用电荷来滤除灰尘和烟灰。在电厂，使用静电过滤器过滤排放到空气中的废气。

[5] Conductors are said to have low resistance to electrical current flow. Conductors usually have three or fewer electrons in the outer orbit of their atoms.

导体对流通的电流阻力小。导体原子的外层轨道通常有三个或少于三个的电子。

[6] The electrons of materials that are insulators are difficult to release. In some insulators, their valence shells are filled with eight electrons. The valence shells of others are over half-filled with electrons. The atoms of materials that are insulators are said to be stable.

绝缘体材料的电子难于释放。在一些绝缘体里，它们的价电子层填充有八个电子。其他绝缘体的价电子层也填充有一半以上的电子。绝缘体材料的原子据说是稳定的。

[7] Semiconductor materials are neither conductors nor insulators. Their classification also depends on the number of electrons that their atoms have in their valence shells. Semiconductors have 4 electrons in their valence shells.

半导体材料既不是导体，也不是绝缘体。它们的分类取决于原子的价电子层电子的数量。半导体原子的价电子层有四个电子。

[8] Static electricity is caused by stationary charges. However, electrical current is the motion of electrical charges from one point to another. Electric current is produced when electrons are removed from their atoms.

静电是由静止的电荷产生的。然而，电流是由电荷从一点到另一点的运动产生的。当电子从它们的原子移除时，就会产生电流。

[9] The movement of valence shell electrons of conductors produces electrical current. The outer electrons of the atoms of a conductor are called free electrons. Energy released by these electrons as they move allows work to be done.

导体价电子层电子的运动产生电流。导体原子的外部电子称为自由电子。当这些电子运动时释放出能量，从而做功。

[10] A circuit is a path or conductor for electric current flow. Electric current flows only when it has a complete, or closed-circuit, path. There must be a source of electrical energy to cause current to flow along a closed path.

电路是电流流通的路径或导体。只有当电路是通路或闭合的路径时才有电流流通。必须有电源来引起电流沿闭合路径流通。

[11] An electrical energy source such as a battery or generator produces current flow through a circuit. As voltage is increased, the amount of current in the circuit is also increased. Voltage is

also called *electromotive force* (*EMF*).

电能源，如电池或发电机，产生流过电路的电流。当电压增加时，电路中电流的数值也增加。电压也称为电动势。

[12] Atoms of some materials give up their free electrons easily. These materials offer low opposition to current flow. Other materials hold their outer electrons and offer high opposition to current flow.

一些材料的原子容易释放它们的自由电子。这些材料对流通电流的阻力小。而其他材料束缚它们的外层电子，对流通电流的阻力大。

[13] Current is measured by connecting the meter into a circuit, i.e. connecting the meter in series with the circuit. Current flows from a voltage source when some device that has resistance is connected to the source.

测量电流时需把仪表接入电路中，即仪表与电路相串联。当某个有阻值的器件与电源相连时就会从电压源产生电流。

[14] There are three types of electrical circuits. They are called series circuits, parallel circuits, and combination circuits. The easiest type of circuit to understand is the series circuit.

电路可分为三种类型，分别为串联电路、并联电路和串并联组合电路。串联电路是最容易理解的电路类型。

[15] For example, the most important AC value is called the effective, or measured, value. This value is less than the peak positive value. A common AC voltage is 220 V, which is used in homes. This is an effective value voltage. Its peak value is about 311V.

例如，最重要的交流值称为有效值或测量值。这个值比正向峰值小。通常应用于我们家庭的交流电压为220V，这个电压是有效值电压。它的峰值电压大约是311V。

[16] The term *phase* refers to time, or the difference between one point and another. If two sine-wave voltages reach their zero and maximum values at the same time, they are in phase.

术语"相位"反映的是某一时间上两点之间的不同。如果两个正弦波在同一时刻到达它们电压的零点和最大值，就说它们是同相位的。

[17] Three-phase AC is used to power large equipment in industry and commercial buildings. It is not distributed to homes. There are three power lines on a three-phase system. Some three-phase systems have a neutral connection and others do not.

在大功率工业设备和商业建筑上使用三相交流。三相交流电不配送给家庭。三相系统有三条动力线。其中一些三相系统有中性线，而其他的没有中性线。

[18] The power ratings of motors and generators are greater when three-phase AC powers are used. For a certain frame size, the rating of a three-phase AC motor is almost 50% larger than a similar single-phase AC motor.

当使用三相交流电源时，电动机和发电机的功率额定值要比单相的大得多。对某一结构尺寸的交流电动机，三相交流电动机的功率额定值要比相似的单相交流电动机大50%。

Chapter 2 Analog Electronics

2.1 Introduction

Electronic systems based on analog principles form an important class of electronic devices. The frequency-modulated (FM) receiver is one common example of analog electronic systems. Although modern receivers contain certainly digital components as well, the transmitted signal is analog. The input to an FM receiver is an FM signal. The signal is an analog, that is, it is a continuous function of time and can have any amplitude. Many electrical instruments, e. g. voltmeters, ammeters, wattmeter and oscilloscopes also utilize analog techniques, at least in part. [1]

Feedback is an important concept in analog electronics. Feedback is a technique by which gain of the analog system can be exchanged for other desirable qualities such as widerbandwidth and linearity. [2] Without feedback, the analog system such as FM or TV receiver would at best offer poor performance.

Operational amplifier is one of the important components in analog electronics. The basic building blocks for an analog circuit are provided by the operational amplifier. [3] In this chapter, we will present some of the common applications of operational amplifiers. Understanding of the benefits of feedback provides the foundation for appreciating the many uses of op-amps in analog electronics.

2.2 Operational Amplifiers

2.2.1 Integrated-Circuit Amplifying Systems

A number of manufacturers market amplifying systems in integrated-circuit packages. A wide range of different intergrated circuits (ICs) are available. Some of these are designed to perform one specific system function. This includes such things as preamplifiers, linear signal amplifiers, and power amplifiers. [4] Other ICs may perform as a complete system. Electrical power for operation and only a limited number of external components are needed to complete the system. Low-power audio amplifiers and stereo amplifiers are examples of these devices.

An operational amplifier or op-amp is a high-performance, directly coupled amplifier circuit containing several transistor devices. [5] The entire assembly is built on a small silicon substrate and packaged as an IC. ICs of this type are capable of high-gain signal amplification from DC to several million hertz. An op-amp is a modular, multistage amplifying device. Operational amplifiers provide basic building blocks for analog circuits in the same way that NOR and NAND gates are basic building blocks for digital circuit.

2.2.2 Properties and Construction

Operational amplifiers have several important properties. They have open-loop gain capabilities in the range of 200, 000 with an input impedance of approximately $2\ M\Omega$. The output impedance is rather low, with values in the range of 50 Ω or less. Their bandwidth, or ability to amplify different frequencies, is rather good. The gain does, however, have a tendency to drop, or roll off, as the frequency increases.

The internal construction of an op-amp is quite complex and usually contains a large number of discrete components. The internal circuitry of an op-amp can be divided into three functional units. Fig. 2-1 shows a simplified diagram of the internal functions of an op-amp. Notice that each function is enclosed in a triangle. Electronic schematics use the triangle to denote the amplification function. This diagram shows that the op-amp has three basic amplification functions. These functions are generally called stages of amplification. A stage of amplification contains one or more active devices and all the associated components is needed to achieve amplification.[6]

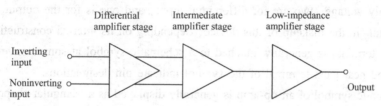

Fig. 2-1 Simplified Diagram of the Internal Functions of an Op-Amp

The first stage, or input, of an op-amp is usually a differential amplifier. This amplifier has two inputs, which are labeled V1 and V2. It provides high gain of the signal difference supplied to the two inputs and low gain for common signals applied to both inputs simultaneously. The input impedance is high to any applied signal. The output of the amplifier is generally two signals of equal amplitude and 180° out of phase. This could be described as a push-pull input and output.

One or more intermediate stages of amplification follow the differential amplifier. An op-amp with only one intermediate stage is shown in Fig. 2-1. This amplifier is designed to shift the operating point to a zero level at the output and has high current and voltage gain capabilities. Increased gain is needed to drive the output stage without loading down the input. The intermediate stage generally has two inputs and a single-ended output.

The output stage of an op-amp has a rather low output impedance and is responsible for developing the current needed to drive an external load. Its input impedance must be great enough that it does not load down the output of the intermediate amplifier.[7] The output stage can be emitter-follower amplifier or two transistors connected in a complementary-symmetry configuration. Voltage gain is rather low in this stage with a sizable amount of current gain.

The audio amplifier, LM379, is a dual 6W audio amplifier. This particular IC offers high-quality performance for stereo phonographs, tape players, recorders, and AM-FM stereo receivers. The internal circuitry of this IC has 52 transistors, 48 resistors, a zener diode, and two silicon diodes. This particular amplifier is designed to operate with a minimum of external components. It contains internal bias regulation circuitry for each amplifier section. Overload protection consists of internal current-limiting and thermal shutting down circuits.[8] The LM 379 has high gain capabilities. The supply voltage can be from 10 V to 35 V. Approximately 0.5 A of current is needed for each amplifier. This particular IC must be mounted to a heat sink. A hole is provided in the housing for a bolted connection to the heat sink. A person working, with an op-amp does not ordinarily need to be concerned with its internal construction. It is helpful, however, to have some general understanding of what the internal circuitry accomplishes. This permits the user to see how the device performs and indicates some of its limitations as a functioning unit. A number of companies manufacture amplifying system ICs today. As a general rule, a person should review the technical data developed by these manufacturers before attempting to use a device.

2.2.3 Schematic Symbol of an OP-AMP

An op-amp has at least five terminals or connections in its construction. Two of these are for the power supply voltage, two are for differential input, and one is for the output.[9] There may be other terminals in the makeup of this device, depending on its internal construction or intended function. Each terminal is generally attached to a schematic symbol at some convenient location. Numbers located near each terminal of the symbol indicate pin designations.

The schematic symbol of an op-amp is generally displayed as a triangular-shaped wedge. The triangle symbol in this case denotes the amplification function. A typical symbol with its terminals labeled is shown in Fig. 2-2. The point, or apex, of the triangle identifies the output. The two leads labeled minus (−) and plus (+) identify the differential input terminals. The minus sign indicates inverting input and the plus sign denotes non-inverting input. A signal applied to the minus input will be inverted 180° at the output. Standard op-amp symbols usually have the inverting input located in the upper-left corner. A signal applied to the plus input will not be inverted and remains in phase with the input.[10] The plus input is located in the lower-left corner of the symbol. In all cases, the two inputs are clearly identified as plus and minus inside the triangle symbol.

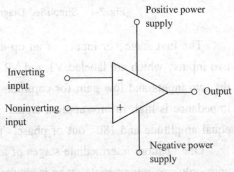

Fig. 2-2 Schematic Symbol of an Op-Amp

Connections or terminals on the sides of the triangle symbol are used to identify a variety of functions. The most significant of these are the two power supply terminals. Normally, the positive voltage terminal (V +) is positioned on the top side and the negative voltage (V −) is on the bottom side. In practice, most op-amps are supplied by a split, or divided, power supply. This supply has V + , ground, and V − terminals. It is important that the correct voltage polarity be supplied to the appropriate terminals, or the device may be permanently damaged. A good rule to follow for most op-amps is not to connect the ground lead of the power supply to V −. An exception to this rule is the current-differencing amplifier (CDA) op-amp. These op-amps are made to be compatible with digital logic ICs and are supplied by a straight 5 − V voltage source.

In the actual schematic symbol of an op-amp, the terminals are generally numbered and the element names are omitted. The manufacturer of the device supplies information sheets that identify the pin-out and operating data. [11]

The schematic symbol of AD741 connected as an open-loop amplifier is shown in Fig. 2-3. The inputs of an op-amp are labeled minus (−) for inverting and plus (+) for the noninverting function, which also denotes the fact that the inputs are differentially related. Essentially, this means that the polarity of the developed output is based on the voltage difference between the two inputs. In Fig. 2-3a, the output voltage is positive with respect to ground. This is the result of the inverting input being made negative with respect to the noninverting input. Reversing the input voltage, as in Fig. 2-3b, causes the output to be negative with respect to ground. For this connection, the noninverting input is negative when the inverting input is positive. In some applications one input may be grounded; the difference voltage is then made with respect to ground. The inputs of Fig. 2-3 are connected in a floating configuration that does not employ a ground. For either type of input, the resulting output polarity is always based on the voltage difference. This characteristic of the input makes the op-amp become an extremely versatile amplifying device.

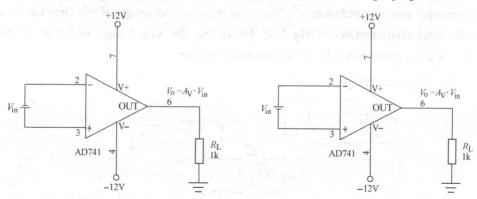

Fig. 2-3 Inverting DC Amplifiers

2.3 Differential Amplifiers

2.3.1 Simple Differential Amplifier

A simple differential amplifier is shown in Fig. 2-4. Assuming an ideal op-amp and that $R_4/R_3 = R_2/R_1$, the output voltage is a constant times the differential input signal, $(V_1 - V_2)$. The gain for the common-mode signal is zero. To minimize

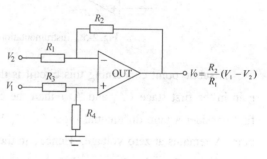

Fig. 2-4 Differential Amplifier

the effects of bias current, we should choose $R_2 = R_4$ and $R_1 = R_3$.

The output impedance of the circuit is zero. The input impedance for the V_1 source is $R_3 + R_4$.

A current that depends on V_1 flows back through the feedback network (R_1 and R_2) into the input source V_2. Thus, as seen by the V_2 source, the circuit does not appear to be passive. Hence, the concept of input impedance does not apply for the V_2 source (unless V_1 is zero).

In some applications, the signal sources contain internal impedances, and the desired signal is the difference between the internal source voltages. Then, we could design the circuit by including the internal source resistances of V_2 and V_1 as part of R_1 and R_3, respectively. However, to obtain very high common-mode rejection, it is necessary to match the ratios of the resistances closely. This can be troublesome if the source impedances are not small enough to be neglected and are not predictable.

2.3.2 Instrumentation-Quality Differential Amplifier

An improved differential amplifier circuit for which the common-mode rejection ratio is not dependent on the internal resistances of the sources is shown in Fig. 2-5. Because of the summing-point constraint at the inputs of A_1 and A_2, the currents drawn from the signal sources are zero. Hence, the input impedances seen by both sources are infinite, and the output voltage is unaffected by the internal source impedances.[12] This is an important advantage of this circuit compared to the simpler differential amplifier of Fig. 2-4. Notice that the second stage of the instrumentation amplifier is a unity-gain version of the differential amplifier.

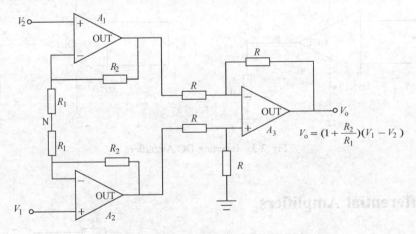

Fig. 2-5 Instrumentation-Quality Differential Amplifier

A subtle point concerning this circuit is that the differential-mode signal experiences a higher gain in the first stage (A_1 and A_2) than the common-mode signal does. To illustrate this point, first consider a pure differential input (i.e., $V_1 = V_2$). Then, because the circuit is symmetrical, point N remains at zero voltage. Hence, in the analysis for a purely differential input signal, point N can be considered to be grounded. In this case, the input amplifiers A_1 and A_2 are configured as non-inverting amplifiers having gains of $(1 + R_2/R_1)$. The differential gain of the second stage is unity. Thus, the overall gain for the differential signal is $(1 + R_2/R_1)$.

Now, consider a pure common-mode signal (i. e. $V_1 = V_2 = V_{cm}$). Because of the summing-point constraint, the voltage between the input terminals of A_1 (or A_2) is zero. Thus, the voltages at the inverting input terminals of A_1 and A_2 are both equal to V_{cm}. Hence, the voltage across the series-connected R_1 resistors is zero, and no current flows through the R_1 resistors. Therefore, no current flows through the R_2 resistors. Thus, the output voltages of A_1 and A_2 are equal to V_{cm}, and we have shown that the first-stage gain is unity for the common-mode signal. On the other hand, the differential gain of the first stage is $(1 + R_2/R_1)$, which can be much larger than unity, thereby achieving a reduction of the common-mode signal amplitude relative to the differential signal. [Notice that if point N were actually grounded, the gain for the common-mode signal would be the same as for the differential signal, namely $(1 + R_2/R_1)$.]

In practice, the series combination of the two R_1 resistors is implemented by a single resistor (equal in value to $2R_1$) because it is not necessary to have access to point N. Thus, matching of component values for R_1 is not required. Furthermore, it can be shown that close matching of the R_2 resistors is not required to achieve a higher differential gain than common-mode gain in the first stage. Since the first stage reduces the relative amplitude of the common-mode signal, matching of the resistors in the second stage is not as critical.

Thus, although it is more complex, the differential amplifier of Fig. 2-5 has better performance than that of Fig. 2-4. Specifically, the common-mode rejection ratio is independent of the internal source resistances, the input impedance seen by both sources is infinite, and resistor matching is not as critical. [13]

2.4 Integrator and Differentiator

2.4.1 Integrator Circuit

The diagram of an integrator is shown in Fig. 2-6, which is a circuit that produces an output voltage proportional to the running-time integral of the input voltage. (By the term running time integral, we mean that the upper limit of integration is t.)

The integrator circuit is often useful in instrumentation applications. For example, consider a signal from an accelerometer that is proportional to acceleration. [14] By integrating the acceleration signal, we obtain a signal proportional to velocity. Another integration yields a signal proportional to position.

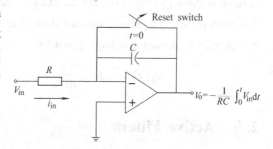

Fig. 2-6 Schematic Diagram of an Integrator

In Fig. 2-6, negative feedback occurs through the capacitor. Thus, assuming an ideal op-amp, the voltage at the inverting op-amp input is zero. The input current is given by

$$i_{in} = \frac{V_{in}}{R} \tag{2-1}$$

The current flowing into the input terminal of the (ideal) op-amp is zero. Therefore, the input current i_{in} flows through the capacitor. We assume that the reset switch is opened at $t=0$. Therefore, the capacitor voltage is zero at $t=0$. The voltage across the capacitor is given by

$$V_C(t) = \frac{1}{C}\int_0^t i_{in}(t)\,dt \tag{2-2}$$

Writing a voltage equation from the output terminal through the capacitor and then to ground through the op-amp input terminals, we obtain

$$V_o(t) = -V_C(t) \tag{2-3}$$

Using Equation (2-1) to substitute into Equation (2-2) and the result into Equation (2-3), we obtain

$$V_o(t) = -\frac{1}{RC}\int_0^t V_{in}\,dt \tag{2-4}$$

Thus, the output voltage is $-1/(RC)$ times the running integral of the input voltage. If an integrator having positive gain is desired, we can cascade the integrator with an inverting amplifier. The magnitude of the gain can be adjusted by the choice of R and C.

Of course, in selecting a capacitor, we usually want to use as small a value as possible to minimize cost, volume, and mass. However, for a given gain constant $(1/(RC))$, smaller C leads to larger R and smaller values of i_{in}. Therefore, the bias current of the op-amp becomes more significant as the capacitance becomes smaller. As usual, we try to design for the best compromise.

2.4.2 Differentiator Circuit

A differentiator circuit that produces an output voltage proportional to the time derivative of the input voltage is shown in Fig. 2-7. By an analysis similar to that used for the integrator, we can show that the circuit produces an output voltage given by

$$V_o(t) = -RC\frac{dV_{in}}{dt} \tag{2-5}$$

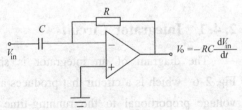

Fig. 2-7 Schematic Diagram of a Differentiator

2.5 Active Filters

A few examples of passive filter design had been considered. In this section, we show how to design lowpass filters composed of resistors, capacitors, and op-amps. Because of the op-amp, these circuits are said to be active filters.[15] In many respects, active filters have improved performance compared to passive circuits.

Active filters have been studied extensively and many useful circuits have been found. Various circuits have been described in the literature that meet these goals to varying degrees. Many complete books have been written that deal exclusively with active filters.[16] In this section, we confine our attention to a particular (but practical) means for implementing lowpass filters.

The magnitude of the Butterworth transfer function is given by

$$|H(f)| = \frac{H_0}{\sqrt{1 + (f/f_B)^{2n}}} \qquad (2\text{-}6)$$

in which the integer n is the order of the filter and f_B is the 3 dB cut-off frequency. Substituting $f=0$ yields $|H(0)| = H_0|$; thus, H_0 is the DC gain magnitude. Plots of this transfer function are shown in Fig. 2-8. Notice that as the order of the filter increases, the transfer function approaches that of an ideal lowpass filter.

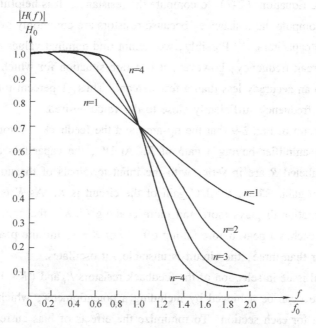

Fig. 2-8 Transfer-Function Magnitudes Versus Frequency for Lowpass Butterworth Filter

An active lowpass Butterworth filter can be implemented by cascading modified Sallen-Key circuits, one of which is shown in Fig. 2-9. In this version of the Sallen-Key circuit, the resistors labeled R have equal values. Similarly, the capacitors labeled C have equal values. Useful circuits having unequal components are possible, but equal components are convenient.

The Sallen-Key circuit shown in Fig. 2-9 is a second-order lowpass filter. To

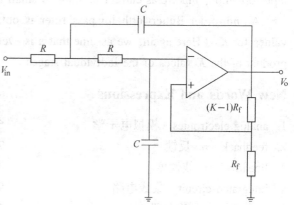

Fig. 2-9 Sallen-Key Lowpass Filter

obtain an nth-order filter, $n/2$ circuits must be cascaded. (We assume that n is even.)

The 3 dB cut-off frequency of the overall filter is related to R and C by

$$f_B = \frac{1}{2\pi RC} \tag{2-7}$$

Usually, we wish to design for a given cut-off frequency. We try to select small capacitance values because this leads to small physical size and low cost. However, Equation (2-7) shows that as the capacitances become small, the resistance values become larger (for a given cut-off frequency). If the capacitance is selected too small, the resistance becomes unrealistically large. Furthermore, stray wiring capacitance can easily affect a high-impedance circuit. Thus, we select a capacitance value that is small, but not too small (say not smaller than 1,000 pF).

In selecting the capacitor, we should select a value that is readily available the tolerance required. Then, we use Equation (2-7) to compute the resistance. It is helpful to select the capacitance first and then compute the resistance, because resistors are commonly available in more finely spaced values than capacitors. [17] Possibly, we cannot find nominal values of R and C that yield exactly the desired break frequency; however, it is a rare situation for which the break frequency must be controlled to an accuracy less than a few percent. Thus, 1 percent tolerance resistors usually result in a break frequency sufficiently close to the value desired.

Notice in the circuit of Fig. 2-9 that the op-amp and the feedback resistors R_f and $(K-1)R_f$ form a non-inverting amplifier having a gain of K. At DC, the capacitors act as open circuits. Then, the resistors labeled R are in series with the input terminals of the non-inverting amplifier and have no effect on gain. Thus, the DC gain of the circuit is K. As K is varied from zero to three, the transfer function displays more and more peaking (i.e., the gain magnitude increases with frequency and reaches a peak before falling off). For $K=3$, infinite peaking occurs. It turns out that for K greater than three, the circuit is unstable – it oscillates.

The most critical issue in selection of the feedback resistors R_f and $(K-1)R_f$ is their ratio. If desired, a precise ratio can be achieved by including a potentiometer, which is adjusted to yield the required DC gain for each section. To minimize the effects of bias current, we should select values such that the parallel combination of R_f and $(K-1)R_f$ is equal to $2R$. However, with FET input op-amps, input bias current is often so small that this is not necessary.

An nth-order Butterworth lowpass filter is obtained by cascading $n/2$ stages having proper values for K. (Here again, we assume that n is even.) The DC gain H_0 of the overall filter is the product of the K values of the individual stages.

New Words and Expressions

1. analog electronics   模拟电子学
2. feedback   n. 反馈
3. linearity   n. 线性度
4. integrated-circuit   集成电路
5. amplifier   n. 放大器
6. component   n. 元件
7. couple   v. 耦合
8. transistor   n. 晶体管
9. silicon   n. 硅
10. package   v. 封装

11. gain   *n.* 增益
12. capability   *n.* 性能
13. impedance   *n.* 阻抗
14. bandwidth   *n.* 带宽
15. label   *v.* 为……标号
16. resistor   *n.* 电阻器
17. zener diode   齐纳二极管
18. heat sink   散热器
19. terminal   *n.* 接线端子
20. power supply   电源
21. schematic   *n.* 原理图
22. minus   *n.* 减号，负号
23. plus   *n.* 加号，正号
24. inverting amplifier   反向放大器
25. non-inverting amplifier   同向放大器
26. differential   *adj.* 差动的，微分的
27. rejection   *n.* 抑制，衰减
28. integrator   *n.* 积分器
29. differentiator   *n.* 微分器
30. capacitor   *n.* 电容器
31. constant   *n.* 常数
32. bias   *n.* 偏置，偏压
33. passive filter   *n.* 无源滤波器
34. active filter   *n.* 有源滤波器
35. implement   *v.* 实现
36. magnitude   *n.* 大小，幅度
37. cascade   *v.* 级联
38. cut-off   *adj.* 截止的

Notes

[1] The signal is an analog, that is, it is a continuous function of time and can have any amplitude. Many electrical instruments, e.g. voltmeters, ammeters, wattmeter and oscilloscopes also utilize analog techniques, at least in part.

信号是模拟信号，也就是说它是时间的连续函数，能有任何幅度。许多电子仪器，如电压表、安培表、功率表以及示波器也使用模拟技术，至少在部分上使用了模拟技术。

[2] Feedback is a technique by which gain of the analog system can be exchanged for other desirable qualities such as wider bandwidth and linearity.

反馈是一项技术，通过反馈模拟系统的增益能被转换为其他所期望的性能，如更高的带宽及线性度。

[3] Operational amplifier is one of the important components in analog electronics. The basic building blocks for an analog circuit are provided by the operational amplifier.

在模拟电子学中，运算放大器是最重要的器件之一。运算放大器是构成模拟电路的基本结构框架。

[4] A number of manufacturers market amplifying systems in integrated-circuit packages. A wide range of different ICs are available. Some of these are designed to perform one specific system function. This includes such things as preamplifiers, linear signal amplifiers, and power amplifiers.

许多制造商以集成电路封装的形式销售放大系统。有大量的不同的集成电路可以使用。其中一些被设计用来执行一些特殊的系统函数，它包括诸如前置放大器、线性信号放大器和功率放大器。

[5] An operational amplifier or op-amp is a high-performance, directly coupled amplifier circuit containing several transistor devices.

运算放大器或称运放是包含若干晶体管器件的高性能、直接耦合的放大器电路。

[6] This diagram shows that the op-amp has three basic amplification functions. These functions are generally called stages of amplification. A stage of amplification contains one or more active devices and all the associated components is needed to achieve amplification.

如图所示的运放有三个基本的放大功能。这些功能通常被称为放大级。一个放大级包含一个或多个有源器件，并且需要所有这些相互关联的元件用以实现放大的性能。

[7] The output stage of an op-amp has a rather low output impedance and is responsible for developing the current needed to drive an external load. Its input impedance must be great enough that it does not load down the output of the intermediate amplifier.

运放的输出级有相当低的输出阻抗，担负着产生驱动外部负载所需电流的责任。它的输入阻抗极高，这样不至于使中间放大器的输出下降。

[8] This particular amplifier is designed to operate with a minimum of external components. It contains internal bias regulation circuitry for each amplifier section. Overload protection consists of internal current-limiting and thermal shutting down circuits.

这种特殊的放大器被设计在带有最少量的外部元件条件下工作。放大器的每一个部分都包含内部偏置调整电路。过载保护由内部的限流电路和热关断电路组成。

[9] An op-amp has at least five terminals or connections in its construction. Two of these are for the power supply voltage, two are for differential input, and one is for the output.

一个运放在结构上至少有五个端子或引线。其中两个为电源引线，两个为不同的输入端，一个为输出端。

[10] The minus sign indicates inverting input and the plus sign denotes non-inverting input. A signal applied to the minus input will be inverted 180° at the output. Standard op-amp symbols usually have the inverting input located in the upper-left corner. A signal applied to the plus input will not be inverted and remains in phase with the input.

负号表示反向输入端，正号表示同向输入端。施加于负号输入端的信号将在输出端产生180°的反向输出。标准的运放符号反向输入端通常位于左上角。施加于正号输入端的信号将不被反向，输出与输入保持同相位。

[11] In the actual schematic symbol of an op-amp, the terminals are generally numbered and the element names are omitted. The manufacturer of the device supplies information sheets that identify the pin-out and operating data.

实际运放的原理图符号，引线端子通常删除了要素名而用数字来标定。器件的制造商提供用于识别引出线和工作数据的信息图表。

[12] Because of the summing-point constraint at the inputs of A_1 and A_2, the currents drawn from the signal sources are zero. Hence, the input impedances seen by both sources are infinite, and the output voltage is unaffected by the internal source impedances.

因为求和点被 A_1 和 A_2 的输入束缚，所以来自信号源的电流下降到零。因此，从两个电源看过去的输入阻抗为无穷大。输出电压不受内部电源阻抗的影响。

[13] Specifically, the common-mode rejection ratio is independent of the internal source resistances, the input impedance seen by both sources is infinite, and resistor matching is not as critical.

特别地，共模抑制比独立于内部的电源阻值。从两个电源看过去的输入阻抗是无穷大的。电阻器的匹配已不再是那样关键了。

[14] The integrator circuit is often useful in instrumentation applications. For example, consider a signal from an accelerometer that is proportional to acceleration.

积分器常用于测量设备中。例如，用于测量来自与加速度成正比的加速计的信号。

[15] In this section, we show how to design lowpass filters composed of resistors, capacitors, and op-amps. Because of the op-amp, these circuits are said to be active filters.

本节介绍的是如何设计由电阻器、电容器和运放组成的低通滤波器。由于带有运放，所以这样的电路称为有源滤波器。

[16] Active filters have been studied extensively and many useful circuits have been found. Various circuits have been described in the literature that meet these goals to varying degrees. Many complete books have been written that deal exclusively with active filters.

有源滤波器被广泛地研究，已经发现了许多有用的电路。文献中描述了各式各样的电路以满足变化的阶的目的。专门地涉及有源滤波器的许多成套书籍已经被撰写。

[17] It is helpful to select the capacitance first and then compute the resistance, because resistors are commonly available in more finely spaced values than capacitors.

首先选择电容值，然后计算电阻值，这是有益的。因为一般情况下，电阻器比电容器有更精细的空间值可以利用。

Chapter 3 Digital Electronics

3.1 Introduction

A circuit that employs a numerical signal in its operation is classified as a digital circuit. Computers, pocket calculators, digital instruments, and numerical control (NC) equipment are common applications of digital circuits. [1] Practically unlimited quantities of digital information can be processed in short periods of time electronically. With operational speed being of prime importance in electronics today, digital circuits are used more frequently. [2]

In this chapter, digital circuit applications are discussed. There are many types of digital circuits that have applications in electronics, including logic circuits, flip-flop circuits, counting circuits, and many others. [3] The first sections of this chapter discuss the number systems that are basic to digital circuit understanding. The remainder of the chapter introduces some of the types of digital circuits and explains Boolean algebra as it is applied to logic circuits.

3.2 Digital Number Systems

The most common number system used today is the decimal system, in which 10 digits are used for counting. The number of digits in the system is called its base (or radix). The decimal system, therefore, has a base of 10.

Numbering systems have a place value, which refers to the placement of a digit with respect to others in the counting process. The largest digit that can be used in a specific place or location is determined by the base of the system. [4] In the decimal system the first position to the left of the decimal point is called the units place. Any digit from 0 to 9 can be used in this place. When number values greater than 9 are used, they must be expressed with two or more places. The next position to the left of the units place in a decimal system is the tens place. The number 99 is the largest digital value that can be expressed by two places in the decimal system. Each place added to the left extends the number system by a power of 10.

Any number can be expressed as a sum of weighted place values. The decimal number 2583, for example, is expressed as $(2 \times 1000) + (5 \times 100) + (8 \times 10) + (3 \times 1)$.

The decimal number system is commonly used in our daily lives. Electronically, however, it is rather difficult to use. Each digit of a base 10 system would require a specific value associated with it, so it would not be practical.

3.2.1 Binary Number System

Electronic digital systems are ordinarily the binary type, which has 2 as its base. Only the numbers 0 or 1 are used in the binary system. [5] Electronically, the value of 0 can be associated with a low-voltage value or no voltage. The number 1 can then be associated with a voltage value larger than 0. Binary systems that use these voltage values are said to have positive logic. Negative logic, by comparison, has a voltage assigned to 0 and no voltage value assigned to 1. Positive logic is used in this chapter.

The two operational states of a binary system, 1 and 0, are natural circuit conditions. When a circuit is turned off or has no voltage applied, it is in the off, or 0, state. An electrical circuit that has voltage applied is in the on, or 1, state. By using transistors or ICs, it is possible to change electronically states in less than a microsecond. Electronic devices make it possible to manipulate millions of 0s and is in a second and thus to process information quickly. [6]

The basic principles of numbering used in decimal numbers apply in general to binary numbers. The base of the binary system is 2, meaning that only the digits 0 and 1 are used to express place value. The first place to the left of the binary point, or starting point, represents the units, or is, location. Places to the left of the binary point are the powers of 2. Some of the place values in base 2 are $2^0 = 1$, $2^1 = 2$, $2^2 = 4$, $2^3 = 8$, $2^4 = 16$, $2^5 = 32$, and $2^6 = 64$.

When bases other than 10 are used, the numbers should have a subscript to identify the base used. The number 100_2 is an example.

The number 100_2 (read "one, zero, zero, base 2") is equivalent to 4 in base 10, or 4_{10}. Starting with the first digit to the left of the binary point, this number has value $(0 \times 2^0) + (0 \times 2^1) + (1 \times 2^2)$. In this method of conversion a binary number to an equivalent decimal number, write down the binary number first. Starting at the binary point, indicate the decimal equivalent for each binary place location where a 1 is indicated. For each 0 in the binary number leave a blank space or indicate a 0. [7] Add the place values and then record the decimal equivalent.

The conversion of a decimal number to a binary equivalent is achieved by repetitive steps of division by the number 2. When the quotient is even with no remainder, a 0 is recorded. When the quotient has a remainder, a 1 is recorded. The division process continues until the quotient is 0. The binary equivalent consists of the remainder values in the order last to first.

3.2.2 Binary-Coded Decimal Number System

When large numbers are indicated by binary numbers, they are difficult to use. For this reason, the Binary-Coded Decimal (BCD) method of counting was devised. In this system four binary digits are used to represent each decimal digit. To illustrate this procedure, the number 105, is converted to a BCD number. In binary numbers, $105_{10} = 1000101_2$.

To apply the BCD conversion process, the base 10 number is first divided into digits according to place values. The number 105_{10} gives the digits 1-0-5. Converting each digit to binary gives 0001-0000-0101BCD. Decimal numbers up to 999_{10} may be displayed by this process with only 12 binary numbers. The hyphen between each group of digits is important when displaying BCD numbers.

The largest digit to be displayed by any group of BCD numbers is 9. Six digits of a number-coding group are not used at all in this system. Because of this, the octal (base 8) and the hexadecimal (base 16) systems were devised. Digital circuits process numbers in binary form but usually display them in BCD, octal, or hexadecimal form. [8]

3.2.3 Octal Number System

The octal (base 8) number system is used to process large numbers by digital circuits. The octal system of numbers uses the same basic principles as the decimal and binary systems.

The octal number system has a base of 8. The largest number used in a base 8 system is 7. The place values starting at the left of the octal point are the powers of eight: $8^0 = 1$, $8^1 = 8$, $8^2 = 64$, $8^3 = 512$, $8^4 = 4096$, and so on.

The process of converting an octal number to a decimal number is the same as that used in the binary-to-decimal conversion process. In this method, however, the powers of 8 are used instead of the powers of 2. The number for changing 382_8 to an equivalent decimal is 258_{10}.

Converting an octal number to an equivalent binary number is similar to the BCD conversion process. The octal number is first divided into digits according to place value. Each octal digit is then converted into an equivalent binary number using only three digits.

Converting a decimal number to an octal number is a process of repetitive division by the number 8. After the quotient has been determined, the remainder is brought down as the place value. When the quotient is even with no remainder, a 0 is transferred to the place position. The number for converting 4098_{10} to base 8 is 10002_8.

Converting a binary number to an octal number is an important conversion process of digital circuits. Binary numbers are first processed at a very high speed. An output circuit then accepts this signal and converts it to an octal signal displayed on a readout device.

Assume that the number 110100100_2 is to be changed to an equivalent octal number. The digits must first be divided into groups of three, starting at the octal point. Each binary group is then converted into an equivalent octal number. These numbers are then combined, while remaining in their same respective places, to represent the equivalent octal number.

3.2.4 Hexadecimal Number System

The hexadecimal number system is used in digital systems to process large number values. The base of this system is 16, which means that the largest number used in a place is 15. Digits used by this system are the numbers 0-9 and the letters A-F. The letters A-F are used to denote the digits 10-15, respectively. The place values to the left of the hexadecimal point are the powers of 16: $16^0 = 1$, $16^1 = 16$, $16^2 = 256$, $16^3 = 4096$, $16^4 = 65536$, and so on.

The process of changing a hexadecimal number to a decimal number is similar to that outlined for other conversions. Initially, a hexadecimal number is recorded in proper digital order. The place values, or powers of the base, are then positioned under the respective digits in step 2. In step 3, the value of each digit is recorded. The values in steps 2 and 3 are then multiplied together and added. The sum gives the decimal equivalent value of a hexadecimal number.

The process of changing a hexadecimal number to a binary equivalent is a simple grouping operation. Initially, the hexadecimal number is separated into digits. Each digit is then converted to a binary number using four digits per group. The binary group is combined to form the equivalent binary number.

The conversion of a decimal number to a hexadecimal number is achieved by repetitive division, as with other number systems. In this procedure the division is by 16 and remainders can be as large as 15.

Converting a binary number to a hexadecimal equivalent is the reverse of the hexadecimal to binary process. Initially, the binary number is divided in groups of four digits, starting at the hexadecimal point. Each number group is then converted to a hexadecimal value and combined to form the hexadecimal equivalent number.

3.3 Binary Logic Circuits

In digital circuit-design applications binary signals are far superior to those of the octal, decimal, or hexadecimal systems. Binary signals can be processed very easily through electronic circuitry, since they can be represented by two stable states of operation. These states can be easily defined as on or off, 1 or 0, up or down, voltage or no voltage, right or left, or any other two-condition states. There must be no in-between state.

The symbols used to define the operational state of a binary system are very important. In positive binary logic, the state of voltage, on, true, or a letter designation (such as A) is used to denote the operational state 1. No voltage, off, false, and the letter A are commonly used to denote the 0 condition. A circuit can be set to either state and will remain in that state until it is caused to change conditions.

Any electronic device that can be set in one of two operational states or conditions by an outside signal is said to be bistable. Relays, lamps, switches, transistors, diodes, and ICs may be used for this purpose. [9] A bistable device has the capability of storing one binary digit or bit of information. By using many of these devices, it is possible to build an electronic circuit that will make decisions based upon the applied input signals. The output of this circuit is a decision based upon the operational conditions of the input. Since the application of bistable devices in digital circuits makes logical decisions, they are commonly called binary logic circuits.

If we were to draw a circuit diagram for such a system, including all the resistors, diodes, transistors, and interconnections, we would face an overwhelming task, and an unnecessary one. [10] Anyone who read the circuit diagram would in their mind group the components into standard circuits and think in terms of the "system" functions of the individual gates. For this reason, we design and draw digital circuit with standard logic symbols. Three basic circuits of this type are used to make simple logic decisions. These are the AND circuit, OR circuit, and the NOT circuit. Electronic circuits designed to perform logic functions are called gates. This term refers to the capability of a circuit to pass or block specific digital signals. The logic gate symbols are shown in Fig. 3-1. The small circle at the output of NOT gate indicates the inversion of the signal. Mathematically, this action is described as $A = \bar{B}$. Thus without the small circle, the triangle would represent an amplifier (or buffer) with a gain of unity. An AND gate has two or more inputs and one output. If all inputs are in the 1 state simultaneously, then there will be a 1 at the output. The AND gate in Fig. 3-1 produces only a 1 output when A and B are both 1. Mathematically, this action is described as $A \cdot B = C$. This expression shows the multiplication operation. An OR gate has also two or more inputs and one output. Like the AND gate, each input to the OR gate has two possible states: 1 or 0. The output of OR gate in Fig. 3-1 produces a 1 when either or both inputs are 1. Mathematically, this action is described as $A + B = C$. This expression shows OR addition. This gate is used to make logic decisions of whether or not a 1 appears at either input.

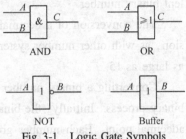

Fig. 3-1 Logic Gate Symbols

An IF-THEN type of sentence is often used to describe the basic operation of a logic state. For example, if the inputs applied to an AND gate are all 1, then the output will be 1. If a 1 is applied to any input of an OR gate, then the output will be 1. If an input is applied to a NOT gate, then the output will be the opposite or inverse. The logic gate symbols in Fig. 3-1 show only the input and output connections. The actual gates, when wired into a digital circuit, would have power supply and grounding connections as well. Fig. 3-2 shows the inner connections of 74LS08, i.e. a quadruple, two-input AND gate chip. Notice that the output power is applied between pin 14 and 7.

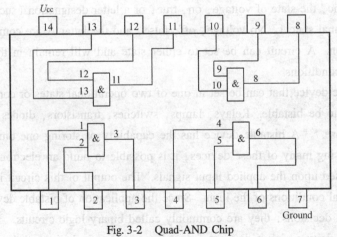

Fig. 3-2 Quad-AND Chip

3.4 Combination Logic Gates

When a NOT gate is combined with an AND gate or an OR gate, it is called a combination logic gate. A NOT-AND gate is called a NAND gate, which is an inverted AND gate. Mathematically, the operation of a NAND gate is $A \cdot B = \overline{C}$. A combination NOT-OR, or NOR, gate produces a negation of the OR function. Mathematically the operation of a NOR gate is $A + B = \overline{C}$. A 1 appears at the output only when A is 0 and B is 0. The logic symbols are shown in Fig. 3-3. The bar over C denotes the inversion, or negative function, of the gate.

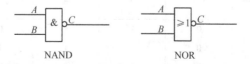

Fig. 3-3 The Basic Combination Logic Gate Symbols

The logic gates discussed here illustrate basic gate operation. In actual digital electronic applications, solid-state components are ordinarily used to accomplish gate functions.

Boolean algebra is a special form of algebra that was designed to show the relationships of logic operations. This form of algebra is ideally suited for analysis and design of binary logic systems. [11] Through the use of Boolean algebra, it is possible to write mathematical expressions that describe specific logic functions. Boolean expressions are more meaningful than complex word statements or elaborate truth tables. The laws that apply to Boolean algebra are used to simplify complex expressions. Through this type of operation, it may be possible to reduce the number of logic gates needed to achieve a specific function before the circuits are designed. [12]

In Boolean algebra the variables of an equation are assigned by letters of the alphabet. Each variable then exists in states of 1 or 0 according to its condition. The 1, or true state, is normally represented by a single letter such as A, B, or C. The opposite state or condition is then described as 0, or false, and is represented by $\overline{A}$, or A'. This is described as NOT A, A negated, or A complemented.

Boolean algebra is somewhat different from conventional algebra with respect to mathematical operations. The Boolean operations are expressed as follows:

Multiplication: A AND B, AB, $A \cdot B$

OR addition: A OR B, $A + B$

Negation, or complementing: NOT A, $\overline{A}$, A'

Assume that a digital logic circuit has three input variables, A, B, and C. The output circuit should operate when only C is on by itself or when A, B, and C are all on expression describes the desired output. Eight (2^3) different combinations of A, B, and C exist in this expression because there are three, inputs. Only two of those combinations should cause a signal that will actuate the output. When a variable is not on (0), it is expressed as a negated letter. The original statement is expressed as follows: With A, B, and C on or with A off, B off, and C on, an output (X) will occur:

$$ABC + \overline{AB}C = X$$

A truth table illustrates if this expression is achieved or not. Table 3-1 shows a truth table for this equation. First, ABC is determined by multiplying the three inputs together. A 1 appears only when the A, B, and C inputs are all 1. Next the negated inputs A and B are determined. Then the products of inputs C, A, and B are listed. The next column shows the addition of ABC and $\overline{A}\,\overline{B}C$. The output of this equation shows that output 1 is produced only when $\overline{A}\,\overline{B}C$ is 1 or when ABC is 1.

Table 3-1 Truth Table for Boolean Equation: $ABC + \overline{A}\,\overline{B}C = X$

| Input | | | ABC | $\overline{A}$ | $\overline{B}$ | $\overline{A}\,\overline{B}C$ | Output |
A	B	C					$ABC + \overline{A}\,\overline{B}C$
0	0	0	0	1	1	0	0
0	0	1	0	1	1	1	1
0	1	0	0	1	0	0	0
0	1	1	0	1	0	0	0
1	0	0	0	0	1	0	0
1	0	1	0	0	1	0	0
1	1	0	0	0	0	0	0
1	1	1	1	0	0	0	1

A logic circuit to accomplish this Boolean expression is shown in Fig. 3-4. Initially the equation is analyzed to determine its primary operational function. Step 1 shows the original equation. The primary function is addition, since it influences all parts of the equation in some way. Step 2 shows the primary function changed to a logic gate diagram. Step 3 shows the branch parts of the equation expressed by logic diagrams, with AND gates used to combine terms. Step 4 completes the process by connecting all inputs together. The circles at inputs $\overline{A}$, $\overline{B}$ of the lower AND gate are used to achieve the negative function of these branch parts.

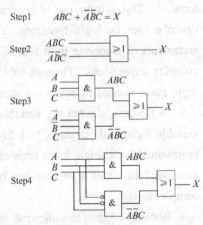

Fig. 3-4 Logic Circuit to Accomplish Boolean Equation $ABC + \overline{A}\,\overline{B}C = X$

The general rules for changing a Boolean equation into a logic circuit diagram are very similar to those outlined. Initially the original equation must be analyzed for its primary mathematical function. This is then changed into a gate diagram that is inputted by branch parts of the equation. Each branch operation is then analyzed and expressed in gate form. The process continues until all branches are completely expressed in diagram form. Common inputs are then connected together.

3.5 Timing and Storage Elements

Digital electronics involves a number of items that are not classified as gates. Circuits or devices of this type have a unique role to play in the operation of a system. Included in this system are such things as timing devices, storage elements, counters, decoders, memory, and registers. Truth tables, symbols, operational characteristics, and applications of these items will be presented here.

Today, these circuits or devices are built primarily on an IC chip. The internal construction of the chip cannot be effectively altered. Operation is controlled by the application of an external signal to the input. As a rule, very little work can be done to control operation other than altering the input signal.

The logic circuits in Fig. 3-4 are combinational circuit because the output responds immediately to the inputs and there is no memory. When memory is a part of a logic circuit, the system is called sequential circuit because its output depends on the input plus its history state. [13]

3.5.1 Flip-Flops

Somebistable multivibrators were already discussed previously. This type of device was used to generate a square wave. It could also be triggered to change states when an input signal is applied. A bistable multivibrator, in the strict sense, is a flip-flop. When it is turned on, it assumes a particular operational state. It does not change states until the input is altered. [14] A flip-flop has two outputs. These are generally labeled Q and $\overline{Q}$. They are always of an opposite polarity. Two inputs are usually needed to alter the state of a flip-flop. A variety of names are used for the inputs. These vary a great deal between different flip-flops.

1. R-S Flip-Flops

Fig. 3-5 shows logic circuit construction of an R-S flip-flop. It is constructed from two NAND gates. The output of each NAND provides one of the inputs for the other NAND. R stands for the reset input and S represents the set input.

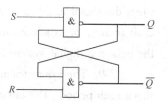

Fig. 3-5 The R-S Flip-Flop Logic Construction

The truth table and logic symbol are shown in Fig. 3-6. Notice that the truth table is somewhat more complex than that of a gate. It shows, for example, the applied input, previous output, and resulting output. To understand the operation of an R-S flip-flop, we must first look at the previous outputs. This is the status of the output before a change is applied to the input. The first four items of the previous outputs are $Q = 1$ and $\overline{Q} = 0$. The second four states have $Q = 0$ and $\overline{Q} = 1$.

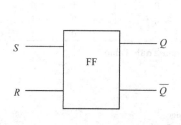

Applied inputs		Previous outputs		Resulting outputs	
S	R	Q	$\overline{Q}$	Q	$\overline{Q}$
0	0	1	0	1/0	0/1
0	1	1	0	1	0
1	0	1	0	0	1
1	1	1	0	1	0
0	0	0	1	0/1	1/0
0	1	0	1	1	0
1	0	0	1	0	1
1	1	0	1	0	1

Fig. 3-6 The R-S Flip-Flop Logic Symbol and Truth Table

Let us consider that R and S are both 1 but that $\overline{Q}$ is 0. In this case of the input to NAND$_S$ is 0 and hence its output, Q, is 1. This is consistent with the assumption that $\overline{Q}$ is 0, which implies that both inputs to NAND_R are 1. By symmetry, the logic circuit will also stable with $Q = 0$ and $\overline{Q} = 1$.

If now R momentarily becomes 0, the output of NAND_R, $\overline{Q}$, will rise to 1, resulting in NANDS having 1 at both inputs. This will force Q to 0, and it will keep $\overline{Q}$ is 1 after R returns to the 1 state. Thus Q is RESET by a 0 at R. Similarly, the SET ($Q = 1$) can be realized by a 0 at S.

The outputs Q and $\overline{Q}$ are unpredictable when the inputs R and S are 0 states. This case is not allowed.

Seldom would individual gates be used to construct a flip-flop, rather than one of the special types for the flip-floppackages on a single chip would be used by a designer.[15]

A variety of different flip-flops are used in digital electronic systems today. In general, each flip-flop type has some unique characteristic to distinguish it from the others. An R-S-T flip-flop, for example, is a triggered R-S flip-flop. It will not change states when the R and S inputs assume a value until a trigger pulse is applied. This would permit a large number of flip-flops to change states all at the same time. Fig. 3-7 shows the logic circuit construction. The truth table and logic symbol are shown in Fig. 3-8. The R and S input are thus active when the signal at the gate input (T) is 1. Normally, such timing, or synchronizing, signals are distributed throughout a digital system by clock pulses, as shown in Fig. 3-9. The symmetrical clock signal provides two

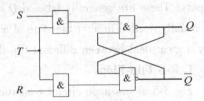

Fig. 3-7 The R-S-T Flip-Flop Logic Construction

times each period. The circuit can be designed to trigger at the leading or trailing edge of the clock. The logic symbols for edge trigger flip-flops are shown in Fig. 3-10.

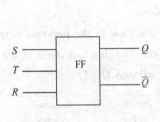

	Applied inputs			Previous outputs		Resulting outputs	
	S	R	T	Q	$\overline{Q}$	Q	$\overline{Q}$
	0	0	0	1	0	1/0	0/1
	0	1	1	1	0	1	0
	1	0	0	1	0	0	1
	1	1	1	1	0	1	0
	0	0	0	0	1	0/1	1/0
	0	1	1	0	1	1	0
	1	0	0	0	1	0	1
	1	1	1	0	1	0	1

Fig. 3-8 The R-S-T Flip-Flop Logic Symbol and Truth Table

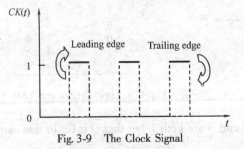

Fig. 3-9 The Clock Signal

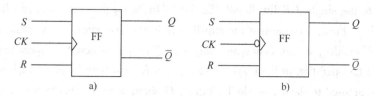

Fig. 3-10 Logic Symbols for Edge Triggered Flip-Flops
a) Leading Edge Triggering b) Trailing Edge Triggering

2. J-K Flip-Flops

Another very important flip-flop has *J-T-K* inputs. A *J-K* flip-flop of this type does not have an unpredictable output state. The *J* and *K* inputs have set and clear input capabilities. These inputs must be present for a short time before the clock or trigger input pulse arrives at *T*. In addition to this, *J-K* flip-flops may employ preset and preclear functions. This is used to establish sequential timing operations. Fig. 3-11 shows the logic symbol and truth table of a *J-K* flip-flop.

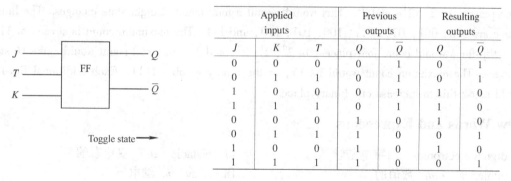

Applied inputs			Previous outputs		Resulting outputs	
J	K	T	Q	$\bar{Q}$	Q	$\bar{Q}$
0	0	0	0	1	0	1
0	1	1	0	1	0	1
1	0	0	0	1	1	0
1	1	1	0	1	1	0
0	0	0	1	0	1	0
0	1	1	1	0	0	1
1	0	0	1	0	1	0
1	1	1	1	0	0	1

Fig. 3-11 The *J-T-K* Flip-Flop Logic Symbol and Truth Table

3.5.2 Counters

A flip-flop has a memory. Each flip-flop can store one binary digit or bit of data. Several flip-flops connected form a counter. Counting is a fundamental digital electronic function.

For an electronic circuit to count, a number of things must be achieved. Basically, the circuit must be supplied with some form of data or information that is suitable for processing. Typically, electrical pulses that turn on and off are applied to the input of a counter. [16] These pulses must initiate a state change in the circuit when they are received. The circuit must also be able to recognize where it is in counting sequence at any particular time. This requires some form of memory. The counter must also be able to respond to the next number in the sequence. In digital electronic systems flip-flops are primarily used to achieve counting. This type of device is capable of changing states when a pulse is applied, has memory, and will generate an output pulse.

There are several types of counters used in digital circuitry today. Probably the most common of these is the binary counter. This particular counter is designed to process two-state or binary information. *J-K* flip-flops are commonly used in binary counters. [17]

Refer now to the single J-K flip-flop of Fig. 3-11. In its toggle state, this flip-flop is capable of achieving counting. First, assume that the flip-flop is in its reset state. This would cause Q to be 0 and $\overline{Q}$ to be 1. Normally, we are concerned only with Q output in counting operations. The flip-flop is now connected for operation in the toggle mode. J and K must both be made high or in the 1 state. When a pulse is applied to the T, or clock, input, Q changes to 1. This means that with one pulse applied, a 1 is generated in the output. The flip-flop has, therefore, counted one time. When the next pulse arrives, Q resets, or changes to 0. Essentially, this means that two input pulses produce only one output pulse. This is a divide-by-two function. For binary numbers, counting is achieved by a number of divide-by-two flip-flops.

To count more than one pulse, additional flip-flops must be employed. For each flip-flop added to the counter, its capacity is increased by the power of 2. With one flip-flop the maximum count was 2^0, or 1. For two flip-flops it would count two places, such as 2^0 and 2^1. This would reach a count of 3 or a binary number of 11. The count would be 00, 01, 10, and 11. The counter would then clear and return to 00. In effect, this counts four state changes. Three flip-flops would count three places, or 2^0, 2^1, and 2^2. This would permit a total count of eight state changes. The binary values are 000, 001, 010, 011, 100, 101, 110, and 111. The maximum count is seven, or 111. Four flip-flops would count four places, or 2^0, 2^1, 2^2, and 2^3. The total count would make 16 state changes. The maximum count would be 15, or the binary number 1111. Each additional flip-flop would cause this to increase one binary place.

New Words and Expressions

1. digital electronics   数字电子学
2. numerical   *adj.* 数值的
3. filp-flop   *n.* 触发器
4. decimal   *n.* 十进制
5. radix   *n.* 底
6. with respect to   相对于
7. power   *n.* 幂
8. weighted   *n.* 权
9. binary   *n.* 二进制
10. manipulate   *v.* 处理，操纵，控制
11. subscript   *n.* 下标，脚标
12. remainder   *n.* 余数
13. quotient   *n.* 商
14. octal   *n.* 八进制
15. hexadecimal   *n.* 十六进制
16. circuitry   *n.* 电路
17. bistable   *adj.* 双稳态的
18. relay   *n.* 继电器
19. capacity   *n.* 性能
20. buffer   *n.* 缓冲器
21. simultaneously   *adv.* 同时地
22. algebra   *n.* 代数
23. truth table   真值表
24. variable   *n.* 变量
25. alphabet   *n.* 字母表
26. complement   *n.* 补码
27. multivibrator   *n.* 多谐振荡器
28. trigger   *v.* 触发
29. symmetrical   *adj.* 对称的
30. leading edge   上升沿
31. trailing (lagging) edge   下降沿
32. counter   *n.* 计数器

Notes

[1] A circuit that employs a numerical signal in its operation is classified as a digital circuit.

Computers, pocket calculators, digital instruments, and numerical control (NC) equipment are common applications of digital circuits.

将数值信号应用于运算过程的电路可以分类为数字电路。计算机、小型计算器、数字仪器以及数控设备通常应用数字电路。

[2] With operational speed being of prime importance in electronics today, digital circuits are used more frequently.

如今,在电子学中,运算速度是最重要的性能之一,因此数字电路被更加频繁地使用。

[3] There are many types of digital circuits that have applications in electronics, including logic circuits, flip-flop circuits, counting circuits, and many others.

有许多类型的数字电路应用在电子学中,它包括逻辑电路、触发电路、计数电路以及许多其他的类型。

[4] Numbering systems have a place value, which refers to the placement of a digit with respect to others in the counting process. The largest digit that can be used in a specific place or location is determined by the base of the system.

编码系统有一个数位值,它是指在计数过程中相对于其他数字的一个数字的位置值。一个特定的数位或位置能使用的最大数字是由该系统的基数确定的。

[5] Electronic digital systems are ordinarily the binary type, which has 2 as its base. Only the numbers 0 or 1 are used in the binary system.

电子数字系统通常指的是二进制系统,该系统以 2 为基数。在二进制系统中只使用 1 和 0 两个数。

[6] By using transistors or ICs, it is possible to change electronically states in less than a microsecond. Electronic devices make it possible to manipulate millions of 0s and is in a second and thus to process information quickly.

通过使用晶体管和集成电路,在不到 1μs 的时间内就可能改变电的运算状态。电子器件使得在 1s 以内操控数以百万的 0 和快速地处理信息成为可能。

[7] Starting at the binary point, indicate the decimal equivalent for each binary place location where a 1 is indicated. For each 0 in the binary number leave a blank space or indicate a 0.

从二进制的小数点开始,当指示为 1 时,对于每一位二进制位置空间给出了十进制等效值。二进制数值中每个 0 保留了一空白空间或指示为 0。

[8] Digital circuits process numbers in binary form but usually display them in BCD, octal, or hexadecimal form.

数字电路以二进制形式处理数码,但通常以 BCD、八进制或十六进制表示。

[9] Any electronic device that can be set in one of two operational states or conditions by an outside signal is said to be bistable. Relays, lamps, switches, transistors, diodes, and ICs may be used for this purpose.

任何电子器件通过外部信号都可以设定在两种运算状态或条件中的一种,这就是说所的双稳态。可以通过使用继电器、灯、开关、晶体管、二极管以及集成电路实现该目的。

[10] If we were to draw a circuit diagram for such a system, including all the resistors, diodes, transistors, and interconnections, we would face an overwhelming task, and an unnecessary

one.

如果所画电路图要包含全部的电阻器、二极管、晶体管以及内部的相互连接，这将是一项巨大的，也是不必要的工作。

[11] Boolean algebra is a special form of algebra that was designed to show the relationships of logic operations. This form of algebra is ideally suited for analysis and design of binary logic systems.

布尔代数是代数的一种特殊形式，被设计用来表示逻辑运算间的关系。这种代数形式完美地适合分析和设计二进制逻辑系统。

[12] The laws that apply to Boolean algebra are used to simplify complex expressions. Through this type of operation, it may be possible to reduce the number of logic gates needed to achieve a specific function before the circuits are designed.

布尔代数法则常常用于简化复杂的表达式。通过这种运算类型，可以减少完成一项特定功能所需的逻辑门的数量。

[13] When memory is a part of a logic circuit, the system is called sequential circuit because its output depends on the input plus its history state.

当存储器作为逻辑电路的一部分时，该系统称为时序电路，时序电路的输出取决于输入以及它的历史状态。

[14] A bistable multivibrator, in the strict sense, is a flip-flop. When it is turned on, it assumes a particular operational state. It does not change states until the input is altered.

严格地讲，多谐振荡器是触发器。假设它开通时处于一种特定的工作状态，这种状态不会改变直到输入信号有改变。

[15] Seldom would individual gates be used to construct a flip-flop, rather than one of the special types for the flip-flop packages on a single chip would be used by a designer.

设计者通常不会使用单独的门来构建触发器，而是会使用封装在一个独立芯片上的特定类型的触发器。

[16] For an electronic circuit to count, a number of things must be achieved. Basically, the circuit must be supplied with some form of data or information that is suitable for processing. Typically, electrical pulses that turn on and off are applied to the input of a counter.

用于计数的电子电路有许多功能必须实现。基本地，必须给电路提供某种形式的适于处理的数据或信息。典型地，开通和关断的电脉冲必须施加于计数器的输入端。

[17] There are several types of counters used in digital circuitry today. Probably the most common of these is the binary counter. This particular counter is designed to process two-state or binary information. J-K flip-flops are commonly used in binary counters.

如今，有几种类型的计数器应用于数字电路中。或许，最常见的就是二进制计数器。这种特殊的计数器被设计用来处理两态或二进位信息。J-K 触发器常常应用于二进制计数器。

Chapter 4 Power Electronics Technology

4.1 Introduction

Power electronics is an enabling technology, providing the needed interface between the electrical source and the electrical load, as depicted in Fig. 4-1. The electrical source and the electrical load can, and often do, differ in frequency, voltage amplitudes and the number of phases. [1] The power electronics interface facilitates the transfer of power from the source to the load by converting voltages and currents from one form to another, in which it is possible for the source and load to reverse roles. [2] The controller shown in Fig. 4-1 allows management of the power transfer process in which the conversion of voltages and currents should be achieved with as high energy-efficiency and high power density as possible. Adjustable-speed electrical drives represent an important application of power electronics.

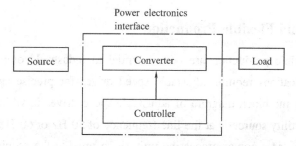

Fig. 4-1 Power Electronics Interface between the Source and the Load

4.2 Applications and the Roles of Power Electronics

Power electronics and drives encompass a wide array of applications. A few important applications and their roles are described below.

4.2.1 Powering the Information Technology

Most of the electronic equipments such as personal computers (PCs) and entertainment systems supplied from the utility mains internally need very low DC voltages. They, therefore, require power converters working in the form of switch-mode for converting the input line voltage into a regulated low DC voltage. [3] Fig. 4-2 shows the distributed architecture typically used in computers in which the incoming AC voltage from the utility is converted into DC voltage, for example, at 24 V. This semi-regulated voltage is distributed within the computer where on-board power supplies in logic-level printed circuit boards convert this 24 V DC input voltage to a lower voltage, for example 5 V DC, which is very tightly regulated. Very large scale integration and higher logic circuitry speed require operating voltages much lower than 5V, hence 3.3 V, 1 V, and eventually, 0.5 V levels would be needed.

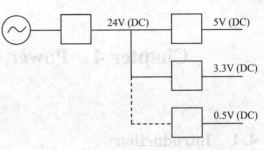

Fig. 4-2　Regulated Low-Voltage DC Power Supplies

Many devices such as cell phones operate from low battery voltages with one or two battery cells as inputs. However, the electronic circuitry within them requires higher voltage, so a circuit is needed to boost the input DC to a higher DC voltage as shown in the block diagram of Fig. 4-3.

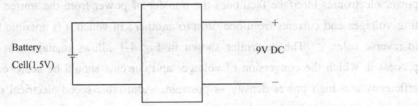

Fig. 4-3　Boost DC-DC Converter Needed in Cell Equipment

4.2.2　Robotics and Flexible Production

Robotics and flexible production are now essential to industrial competitiveness in a global economy. These applications require adjustable speed drives for precise speed and position control.[4] Fig. 4-4 shows the block diagram of adjustable speed drives in which the AC input from a 1-phase or a 3-phase utility source is at the line frequency of 50 Hz or 60 Hz. The role of the power electronics interface, as a power-processing unit, is to provide the required voltage to the motor.[5] In the case of a DC motor, DC voltage of adjustable magnitude is supplied to control the speed of the motor. In the case of an AC motor, the power electronics interface provides sinusoidal AC voltages with adjustable amplitude and frequency to control the speed of the motor.[6] In certain cases, the power electronics interface may be required to allow bi-directional power flow through it, between the utility and the motor-load.

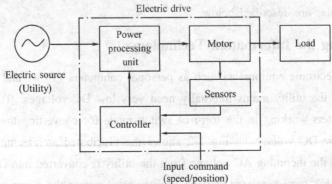

Fig. 4-4　Block Diagram of Adjustable Speed Drives

Induction heating and electrical welding, shown in the block diagrams of the Fig. 4-5 and Fig. 4-6 are other important industrial applications of power electronics for flexible production.

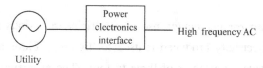

Fig. 4-5 Power Electronics Interface Required for Electric Heating

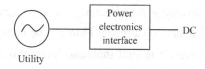

Fig. 4-6 Power Electronics Interface Required for Electrical Welding

4.2.3 Energy and the Environment

There is a growing body of evidence that burning of fossil fuels causes global warming that may lead to disastrous environmental consequences. Power electronics and drives can play a crucial role in minimizing the use of fossil fuels, as briefly discussed below.

4.2.3.1 Energy Conservation

It's an old adage: a penny saved is a penny earned. Not only energy conservation leads to financial savings, it helps the environment. The potentials for energy conservation in some applications are discussed below.

1. Electric-Motor Driven Systems

Electric motors, including their applications in Heating, Ventilating and Air Conditioning (HVAC), are responsible for consuming one-half to two-thirds of all the electricity generated. Traditionally, motor-driven systems run at a nearly constant speed and their output, for example, flow rate in a pump, is controlled by wasting a portion of the input energy across a throttling valve. [7] This waste may be eliminated by efficiently controlling the speed of the pump motor using an adjustable-speed electrical drive, as shown in Fig. 4-7.

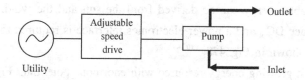

Fig. 4-7 Role of Adjustable Speed Drives in Pump-Driven System

One out of three new homes now uses electrical heat pump, in which adjustable-speed drive can reduce energy consumption by as much as 30 percent by eliminating on-off cycling of the compressor and running the heat pump at a speed that matches the thermal load of the building. The same is true for air conditioners.

A Department of Energy report estimates that electricity annually saved is equivalent to the annual electricity usage of the entire state of New York by operating all these motor-driven systems more efficiently in the United States.

2. Lighting

As shown ina pie chart for percentage use of electricity in various sectors in the U. S. , approximately one-fifth of electricity produced is used for lighting. Fluorescent lights are more efficient than incandescent lights by a factor of three to four. The efficiency of fluorescent lights can be further improved by using high-frequency power electronic ballasts that supply 30 kHz to 40 kHz to the light bulb, further increasing the efficiency by approximately 15 percent. Compared to incandescent light bulbs, high-frequency Compact Fluorescent Lamps (CFLs) improve efficiency by a factor of four, last much longer (several thousand hours more), and their relative cost, although high at present, is dropping. [8]

3. Transportation

Electric drives offer huge potential for energy conservation in transportation. With the progress in battery and fuel cell technologies being reported continually for Electrical Vehicles (EVs), Hybrid Electrical Vehicles (HEVs) are sure to make a huge impact. According to the U. S. Environmental Protection Agency, the estimated gas mileage of the Hybrid Electrical Vehicle in combined city and highway driving is 48 miles per gallon. This is in comparison to the gas mileage of 22. 1 miles per gallon for an average passenger car in the U. S. in year 2001. Since automobiles are estimated to account for about 20% emission of all CO_2 that is a greenhouse gas, doubling the gas mileage of automobiles would have an enormous positive impact.

EVs and HEVs, of course, need power electronicsworking in the form of adjustable-speed electric drives. Even in conventional automobiles, power electronic based load has grown to the extent that it is difficult to supply it from a 12/14 V battery system and there are serious proposals to raise it to 36/42 V level in order to keep the copper bus bars needed to carry currents to a manageable size. Add to automobiles other transportation systems, such as light rail, fly-by-wire planes, all-electrical ships and drive-by-wire automobiles, represent a major application area of power electronics. [9]

4.2.3.2 Renewable Energy

Clean and renewable energy can be derived from the sun and the wind. In photovoltaic systems, solar cells produce DC, and a power electronics interface is required to transfer the power to the utility system, as shown in Fig. 4-8. [10]

Wind is the fastest growing energy resource with enormous potential. Fig. 4-9 shows the need of power electronics in wind-electrical systems as the interface between the variable-frequency AC and the line-frequency AC of the utility grid.

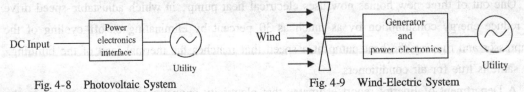

Fig. 4-8 Photovoltaic System   Fig. 4-9 Wind-Electric System

4.2.4 Utility Applications of Power Electronics

Applications of power electronics and electrical drives in power systems are growing rapidly. In distributed generation, power electronics is needed to interface non-conventional energy sources such as wind, photovoltaic, and fuel cells to the utility grid. Use of power electronics allows control the flow of power on transmission lines. It is especially significant in a deregulated utility environment. Also, the security and the efficiency of power systems operation increase the use of power electronics in utility applications.

Uninterruptible Power Supplies (UPSs) are used for critical loads that must not be interrupted during power outage.[11] The power electronics interface for UPS has line-frequency voltages at both ends, although the number of phases may be different, and a means for energy storage is provided usually by batteries, which supply power to the load during the utility outage.

4.2.5 Strategic Space and Defense Applications

Power electronics is essential for space exploration and interplanetary travel. Defense has always been an important application. Power electronics will play a huge role in tanks, ships, and planes in which the replacement of hydraulic drives by electrical drives can offer significant cost, weight and reliability advantages.

4.3 Structure of Power Electronics Interface

By reviewing the role of power electronics in various applications discussed earlier, we can summarize that power electronics interface is needed to efficiently control the transfer of power between DC-DC, DC-AC, and AC-AC systems. In general, the power is supplied by the utility and hence, as depicted by the block diagram of Fig. 4-10, the line-frequency AC is at one end. At the other end, one of the following is synthesized: adjustable magnitude DC, sinusoidal AC of adjustable frequency and amplitude, or high frequency AC as in the case of CFLs and induction heating. Applications that do not require utility interconnection can be considered as the subset of the block diagram shown in Fig. 4-10.

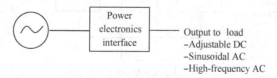

Fig. 4-10 Block Diagram of Power Electronics Interface

4.3.1 Voltage-Link Structure

To provide the needed functionality to the interface in Fig. 4-10, the transistors and diodes, which can only block voltage of one polarity, have led to a commonly used voltage-link structure shown in Fig. 4-11.

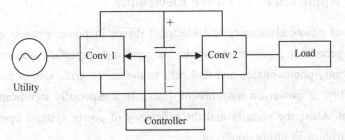

Fig. 4-11 Voltage-Link Structure of Power Electronics Interface

This structure is composed of two separate converters, one on the utility side and the other on the load side. The DC ports of these two converters are connected to each other with a parallel capacitor forming a DC-link, across which the voltage polarity does not reverse, thus allowing unipolar voltage-blocking transistors to be used within these converters.

In the structure of Fig. 4-11, the capacitor in parallel with the two converters forms a DC voltage-link, hence this structure is called a voltage-link (or a voltage-source) structure. This structure is used in a very large power range, from a few tens of watts to several megawatts, even extending to hundreds of megawatts in utility applications. Therefore, we will mainly focus on this voltage-link structure in this book.

4.3.2 Current-Link Structure

At extremely high power levels, usually in utility-related applications, it may be advantageous to use a current-link (also called current-source) structure, where as shown in Fig. 4-12, an inductor in series between the two converters acts as a current-link. These converters generally consist of thyristors and the current in them is "commutated" from one AC phase to another by means of the AC line voltages.

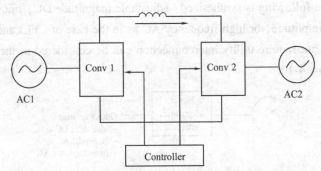

Fig. 4-12 Current-Link Structure of Power Electronics Interface

4.3.3 Matrix Converters (Direct-Link Structure)

Lately in certain applications, a matrix converter structure is being reevaluated, where theoretically there is no energy storage element between the input and the output sides. Therefore, we can consider it to be a direct-link structure where input ports are connected to output ports by switches that can carry currents in both directions when on, and block voltages of either polarity when off.

4.4 Voltage-Link Structure Applications

In the voltage-link structure, the role of the utility-side converter is to convert line-frequency utility voltages to an unregulated DC voltage. This can be done by a diode-rectifier circuit like that discussed in basic electronics courses. [12] At present, we will focus our attention on the load-side converter in the voltage-link structure, where a DC voltage is applied as the input on one end, as shown in Fig. 4-11.

Applications dictate the functionality needed of the load-side converter. Based on the desired output of the converter, we can group these functionalities as follows:

Group 1 Adjustable DC or a low-frequency sinusoidal AC output in
- DC and AC motor drives;
- Uninterruptible power supplies;
- Regulated DC power supplies without electrical isolation;
- Utility – related applications.

Group 2 High-frequency AC in
- Compact fluorescent lamps;
- Induction heating;
- Regulated DC power supplies where the DC output voltage needs to be electrically isolated from the input, and the load-side convener internally produces high-frequency AC, which is passed through a high-frequency transformer and then rectified into DC.

We will discuss converters used in applications belonging to both groups. However, we will begin with converters for Group 1 applications where the load-side voltages are DC or low-frequency AC.

4.4.1 Switch-Mode Conversion: Switching Power-Pole as the Building Block

Achieving high energy-efficiency for applications belonging to either group mentioned above requires switch-mode conversion, where in contrast to linear power electronics, transistors (and diodes) are operated as switches, either on or off.

This switch-mode conversion can be explained by its basic building block, a switching power-pole A, as shown in Fig. 4-13a. It effectively consists of a bi-positional switch, which forms a two-port: a voltage-port across a capacitor with a voltage V_{in} that cannot change instantaneously, and a current-port due the series inductor through which the current cannot change instantaneously. For now, we will assume the switch ideal with two positions: up or down, dictated by a switching signal q_A which takes on two values: 1 and 0, respectively.

The bi-positional switch "chops" the input DC voltage V_{in}, into a train of high-frequency voltage pulses, shown by V_A waveform in Fig. 4-13b, by switching up or down at a high repetition rate, called the switching frequency f_s. Controlling the pulse width within a switching cycle allows control over the switching-cycle-averaged value of the pulsed output, and this pulse-width modulation forms the basis of synthesizing adjustable DC and low-frequency sinusoidal AC outputs, as described in the next section. High-frequency pulses are clearly needed in applications such as compact fluorescent lamps, induction heating, and DC power supplies where electrical isolation is achieved by means of a high-frequency transformer.[13] A switch-mode converter consists of one or more such switching power-poles.

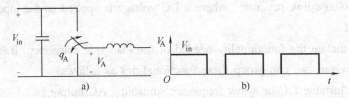

Fig. 4-13 Switching Power-Pole as the Building Block in Converters

4.4.2 Pulse Width Modulation (PWM) of the Switching Power-Pole (constant f_s)

For the applications in Group 1, the objective of the switching power-pole drawn in Fig. 4-13a is to synthesize the output voltage such that its switching-cycle-average is of the desired value: DC or AC that varies sinusoidally at a low-frequency, compared to f_s. Switching at a constant switching-frequency f_s produces a train of voltage pulses in Fig. 4-14 that repeat with a constant switching time-period T_s, equal to $1/f_s$.

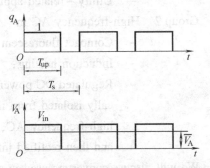

Fig. 4-14 PWM of the switching power-pole

Within each switching cycle with the time-period T_s ($= 1/f_s$) in Fig. 4-14, the switching-cycle-averaged value $\bar{v}_A$ of the waveform is controlled by the pulse width T_{up} (during which the switch is in the up position and v_A equals V_{in}), as a ratio of T_s:

$$\bar{v}_A = \frac{T_{up}}{T_s} V_{in} \quad 0 \leq d_A \leq 1 \tag{4-1}$$

Where $d_A = T_{up}/T_s$, which is the average of the q_A waveform as shown in Fig. 4-14, is defined as the duty-ratio of the switching power-pole A, and the switching-cycle-averaged voltage is indicated by a " $-$ " on top. The switching-cycle-averaged voltage and the switch duty-ratio are expressed by lowercase letters since they may vary as functions of time. The control over the switching-cycle-averaged value of the output voltage is achieved by adjusting or modulating the pulse width, which later on will be referred to as Pulse-Width-Modulation (PWM).[14]

4.4.3 Switching Power-Pole in a Buck DC-DC Converter

As an example, we will consider the switching power-pole in a Buck convener to step-down the input DC voltage V_{in}, as shown in Fig. 4-15a, where a capacitor is placed in parallel with the load, to form a low-pass L-C filter with the inductor, to provide a smooth voltage to the load.

In steady state, the DC (average) input to this L-C filter has no attenuation, hence the average output voltage V_o equals the switching-cycle-average, $\overline{V}_A$, of the applied input voltage. Based on Eq. 4-1, by controlling d_A, the output voltage can be controlled in a range from V_{in} down to 0:

$$V_o = \overline{V}_A = d_A V_{in} \quad (0 \leq V_o \leq V_{in}) \tag{4-2}$$

In spite of the pulsating nature of the instantaneous output voltage $V_A(t)$, the series inductance at the current-port of the pole ensures that the current $i_L(t)$ remains relatively smooth, as shown in Fig. 4-15b.

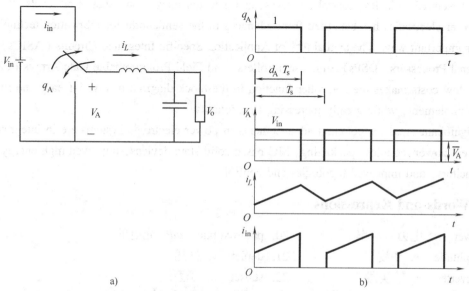

Fig. 4-15 Switching Power-Pole in a Buck Converter

4.5 Recent and Potential Advancements

Given the need in a plethora of applications, the rapid growth of this field is fueled by revolutionary advances in semiconductor fabrication technology. The power electronics interface of Fig. 4-1 consists of solid-state devices, which operate as switches, changing from on to off and vice versa at a high switching frequency. There has been a steady improvement in the voltage and current handling capabilities of solid-state devices such as diodes and transistors, and their switching speeds (from on to off, and vice versa) have increased dramatically, with some devices switching in tens of ns.[15] Devices that can handle voltages in kVs and currents in kAs are now available. Moreover, the costs of these devices are in a steady decline.

These semiconductor switches are integrated in a single package with all the circuitry needed to make them switch, and to provide the necessary protection.[16] A progressive integration, as defined by the concept of Power Electronics Building Block (PEBB) below has numerous benefits such as technology insertion and upgrade via standard interfaces, reduced maintenance via plug and play modules, reduced cost via increased product development efficiency, reduced time to market, reduced commissioning cost, reduced design and development risk, and increased competition in critical technologies. PEBB is a broad concept that incorporates the progressive integration of power devices, gate drives, and other components into building blocks, with clearly defined functionality that provides interface capabilities to serve multiple applications. This building block approach results in reduced cost, losses, weight, size, and engineering effort for the application and maintenance of power electronics systems. Based on the functional specifications of PEBB and the performance requirements of the intended applications, the PEBB designer addresses the details of device stresses, stray inductances, switching speed, losses, thermal management, protection, measurements of required variables, control interfaces, and potential integration issues at all levels.[17]

Power electronics has benefited from advances in the semiconductor fabrication technology in another important way. The availability of Application Specific Integrated Circuits (ASICs), Digital Signal Processors (DSPs), micro-controllers, and Field Programmable Gate Arrays (FPGAs) at very low costs makes the controller function in the block diagram of Fig. 4-1 easy and inexpensive to implement, while greatly increasing functionality.

Significant areas for potential advancements in power electronic systems are in integrated and intelligent power modules, packaging, SIC-based solid-state devices, improved high energy density capacitors, and improved topologies and control.[18]

New Words and Expressions

1. power　　*n.* 电力，功率
2. amplitude　　*n.* 幅度
3. converter　　*n.* 转换器
4. adjustable-speed　　调速
5. encompass　　*v.* 包含，拥有
6. flexible　　*adj.* 柔性的
7. magnitude　　*n.* 量值
8. sinusoidal　　*adj.* 正弦的
9. utility　　*n.* 中心电站
10. welding　　*n.* 焊接
11. ventilate　　*v.* 使通风
12. pump　　*n.* 泵
13. throttle　　*n.* 节流阀
14. compressor　　*n.* 压缩机
15. thermal　　*adj.* 热的
16. fluorescent　　*adj.* 荧光的
17. incandescent　　*adj.* 白炽的
18. mileage　　*n.* 里程
19. renewable　　*adj.* 可恢复的
20. photovoltaic　　*adj.* 光电的
21. depict　　*v.* 描述
22. subset　　*n.* 子系统
23. matrix　　*n.* 矩阵
24. dictate　　*v.* 规定，决定
25. isolation　　*n.* 隔离
26. power supply　　电源
27. transformer　　*n.* 变压器
28. cycle　　*n.* 周期
29. constant　　*n.* 常数
30. modulate　　*v.* 调制
31. filter　　*n.* 滤波器
32. inductor　　*n.* 电感器
33. attenuation　　*n.* 衰减
34. plethora　　*n.* 过多，过剩
35. maintenance　　*n.* 维护，保养
36. module　　*n.* 模块
37. loss　　*n.* 损耗
38. topology　　*n.* 拓扑

Notes

[1] The electrical source and the electrical load can, and often do, differ in frequency, voltage amplitudes and the number of phases.

电源和负载可以且常常在频率、电压幅度及相位值方面不同。

[2] The power electronics interface facilitates the transfer of power from the source to the load by converting voltages and currents from one form to another, in which it is possible for the source and load to reverse roles.

电力电子线路接口方便了功率从电源到负载的传输，它把电压和电流从一种形式转换为另一种形式，从电源到负载其可能扮演着反向的角色。

[3] Most of the electronic equipments such as personal computers (PCs) and entertainment systems supplied from the utility mains internally need very low DC voltages. They, therefore, require power converters working in the form of switch-mode for converting the input line voltage into a regulated low DC voltage.

大部分电子设备，如由电站干线供电的个人计算机及娱乐系统，其内部需要很低的直流电压。因此，它们要求功率转换器工作在开关模式，把输入的电路电压转换为可调整的低的直流电压。

[4] Robotics and flexible production are now essential to industrial competitiveness in a global economy. These applications require adjustable speed drives for precise speed and position control.

在全球经济方面，机器人学和柔性产品对于工业竞争来说是必不可少的。这些应用要求调速驱动用于精确的速度和位置控制。

[5] The role of the power electronics interface, as a power-processing unit, is to provide the required voltage to the motor.

作为功率处理单元，电力电子线路接口的作用是提供电动机工作所需的电压。

[6] In the case of a DC motor, DC voltage of adjustable magnitude is supplied to control the speed of the motor. In the case of an ac motor, the power electronics interface provides sinusoidal ac voltages with adjustable amplitude and frequency to control the speed of the motor.

就直流电动机而言，它提供可调节的直流电压，从而控制电动机的转速。就交流电动机而言，电力电子线路接口提供幅度和频率可调节的正弦电压，从而控制电动机的转速。

[7] Traditionally, motor-driven systems run at a nearly constant speed and their output, for example, flow rate in a pump, is controlled by wasting a portion of the input energy across a throttling valve.

传统的电动机驱动系统运行在几乎恒定的转速条件下。它们的输出，如水泵的流量控制，是通过消耗部分节流阀输入的能量来实现的。

[8] Compared to incandescent light bulbs, high-frequency compact fluorescent lamps (CFLs) improve efficiency by a factor of four, last much longer (several thousand hours more), and their relative cost, although high at present, is dropping.

高频密集型荧光灯与白炽灯相比，效率提升了三倍，寿命更长久（几千小时以上）。虽然它们的相对成本目前高些,但是正在下降。

[9] Add to automobiles other transportation systems, such as light rail, fly-by-wire planes, all-electrical ships and drive-by-wire automobiles, represent a major application area of power electronics.

除了汽车,其他的运输系统,如轻轨、遥控自动驾驶飞机、纯电动轮船以及电动汽车都代表了电力电子学的一个主要应用领域。

[10] Clean and renewable energy can be derived from the sun and the wind. In photovoltaic systems, solar cells produce DC, and a power electronics interface is required to transfer the power to the utility system, as shown in Fig. 4-8.

清洁和可再生的能源可来源于太阳和风。在光伏系统中,太阳能电池产生直流电,需要电力电子线路接口将电能传输到公共电站系统,如图4-8所示。

[11] Uninterruptible Power Supplies (UPSs) are used for critical loads that must not be interrupted during power outage.

不间断电源(UPS)被用于重要的负载,可使负载电源不因电源断电而中断。

[12] In the voltage-link structure, the role of the utility-side converter is to convert line-frequency utility voltages to an unregulated DC voltage. This can be done by a diode-rectifier circuit like that discussed in basic electronics courses.

在压链式结构中,公共电站侧转换器的作用是把线频公共电压转换为可调整的直流电压。如同基础电子学课程所讨论的,这能通过二极管整流器电路实现。

[13] High-frequency pulses are clearly needed in applications such as compact fluorescent lamps, induction heating, and DC power supplies where electrical isolation is achieved by means of a high-frequency transformer.

在诸如密集型荧光灯、感应加热的应用中,以及通过高频变压器实现电气隔离的内部直流电源中,高频脉冲显然是所需的。

[14] The control over the switching-cycle-averaged value of the output voltage is achieved by adjusting or modulating the pulse width, which later on will be referred to as Pulse-Width-Modulation (PWM).

通过调节或调制脉冲的宽度,能实现对输出电压开关周期平均值的控制,后者被称之为脉冲宽度调制。

[15] There has been a steady improvement in the voltage and current handling capabilities of solid-state devices such as diodes and transistors, and their switching speeds (from on to off, and vice versa) have increased dramatically, with some devices switching in tens of ns.

固态器件的电压和电流控制容量已经有一个稳定的进步,如二极管、晶体管以及它们的开关速度(从开到关,反之亦然)已经显著地增加,一些器件的开关时间只有几十纳秒。

[16] These semiconductor switches are integrated in a single package with all the circuitry needed to make them switch, and to provide the necessary protection.

这些半导体开关和需要它们开关及提供必要保护的全部电路一起集成在一个独立的封装里。

[17] Based on the functional specifications of PEBB and the performance requirements of

the intended applications, the PEBB designer addresses the details of device stresses, stray inductances, switching speed, losses, thermal management, protection, measurements of required variables, control interfaces, and potential integration issues at all levels.

基于电力电子构建框架的功能详述，以及所计划应用的性能要求，电力电子构建框架的设计者访问了器件应力、开关速度、损耗、热管理、保护、所要求变量的测量、控制界面以及各种水平的潜在集成出版物的细节。

[18] Significant areas for potential advancements in power electronic systems are in integrated and intelligent power modules, packaging, SIC-based solid-state devices, improved high energy density capacitors, and improved topologies and control.

在电力电子系统中，具有潜在提升的重要领域是集成和智能化的功率模块、封装、基于SIC的固态器件、改进的大能量密度电容器以及改进的拓扑结构与控制。

Chapter 5 Magnetism and Electromagnetism

5.1 Introduction

Magnetism has been studied for many years. Some metals in their natural state attract small pieces of iron. This property is called magnetism. Materials that have this ability are called natural magnets. [1] The first magnets used were called lodestones. Now, artificial magnets are made in many different strengths, sizes, and shapes. Magnetism is important because it is used in electric motors, generators, transformers, relays, and many other electrical devices. The earth itself has a magnetic field like a large magnet.

Electromagnetism is magnetism that is brought about due to electrical current flow. There are many electrical machines that operate because of electromagnetism. [2] This chapter deals with magnetism and electromagnetism and some important applications.

5.2 Permanent Magnets

Magnets are made of iron, cobalt, or nickel materials, usually in an alloy (or mixture). Each end of the magnet is called a pole. If a magnet is broken, each part becomes a magnet with two poles. Magnetic poles always occur in pairs. When a magnet is suspended in air so that it can turn freely, one pole will point to the North Pole of the earth, which explains why compasses can be used to determine direction. The north pole of a magnet attracts the south pole of another magnet. A north pole repels another north pole, and a south pole repels another south pole. The two laws of magnetism are like poles repel, and unlike poles attract.

Some materials retain magnetism longer than others. Hard steel holds its magnetism much longer than soft steel. A magnetic field is set up around any magnetic material. The field is made up of lines of force, or magnetic flux. These magnetic flux lines are invisible. They never cross one another, but they always form individual closed loops around a magnetic material. They have a definite direction from the north to the south pole along the outside of a magnet. [3] When magnetic flux lines are close together, the magnetic field is strong. When magnetic flux lines are farther apart, the field is weaker. [4] The magnetic field is strongest near the poles. Lines of force pass through all materials. It is easy for lines of force to pass through iron and steel. Magnetic flux passes through a piece of iron as shown in Fig. 5-1. This type of permanent magnet is called a bar magnet.

When magnetic flux passes through a piece of iron, the iron acts like a magnet. Magnetic poles are formed due to the influence of the flux lines. These are called induced poles. The induced poles and the magnet's poles attract and repel each other. Magnets attract pieces of soft iron in this way. It is possible to magnetize pieces of metal temporarily by using a bar magnet. If a magnet is passed over the top of a piece of iron several times in the same direction, the soft iron becomes magnetized and stays magnetized for a short period of time. [5]

When a compass is brought near the north pole of a magnet, the north-seeking pole of the compass is attracted to it. The polarities of the magnet may be determined by observing a compass brought near each pole. Compasses detect the presence of magnetic fields.

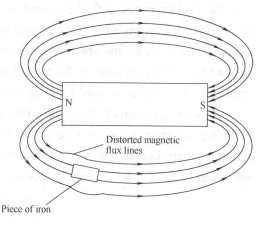

Fig. 5-1　Magnetic Flux Lines Distorted Through a Piece of Iron

Horseshoe magnets are similar to bar magnets. They are bent in the shape of a horseshoe, as shown in Fig. 5-2. This shape gives more magnetic field strength than a similar bar magnet, since the magnetic poles are closer. The magnetic field strength is more concentrated into one area. Many electrical devices use horseshoe magnets.

A magnetic material can lose some of its magnetism if it is jarred or heated. People must be careful when handling equipment that contains permanent magnets. [6] A magnet also becomes weak-

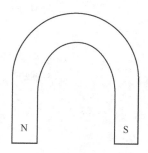

Fig. 5-2　Horseshoe Magnet

ened by loss of magnetic flux. Magnets should always be stored with a keeper, which is a soft-iron piece used to join magnetic poles. The keeper provides the magnetic flux with all easy paths between poles. The magnet will retain its greatest strength for a longer period of time if keepers are used. Bar magnets should always be stored in pairs with a north pole and a south pole placed together. A complete path for magnetic flux is made in this way.

5.3　Magnetic Field around Conductors and a Coil

5.3.1　Magnetic Field around Conductors

Current-carrying conductors produce a magnetic field. A compass is used to show that the magnetic flux lines are circular in shape. The conductor isin the center of the circular shape. The direction of the current flow and the magnetic flux lines can be shown by using the left-hand rule of magnetic flux. [7] A conductor is held in the left hand, as shown in Fig5-3a. The thumb points in the direction of current flow from negative to positive. The fingers then encircle the conductor in the direction of the magnetic flux lines.

The circular magnetic field produced around a conductor is stronger near tile conductor and becomes weaker farther away from the conductor. A cross-sectional end view of a conductor with current flowing toward tile observer is shown in Fig. 5-3b. Current flow toward the observer is shown by a circle with a dot in the center. Notice that the direction of the magnetic flux lines is clockwise. This can be verified by using the left-hand rule.

When the direction of current flow through a conductor is reversed, the direction of the magnetic lines of force is also reversed. The cross-sectional end view of a conductor in Fig. 5-3c shows a current flow in a direction away from the observer. Notice that the direction of the magnetic lines of force is now counterclockwise.

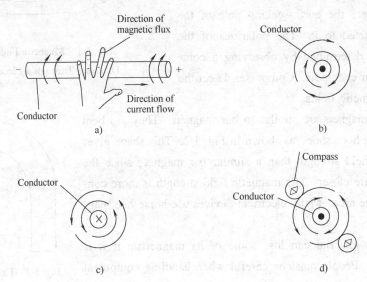

Fig. 5-3 The Direction Relationship of Current Flow and Magnetic Flux

The presence of magnetic lines of force around a current-carrying conductor can be observed by using a compass. When a compass is moved around the outside of a conductor, the needle aligns itself tangent to the lines of force, as shown in Fig. 5-3d. The needle does not point toward the conductor. When current flows in the opposite direction, the compass polarities reverse. The compass needle aligns itself tangent to the conductor.

5.3.2 Magnetic Field around a Coil

The magnetic field around one loop of wire is shown in Fig. 5-4. Magnetic flux lines extend around the conductor as shown. Inside the loop, the magnetic flux is in one direction, i.e. a direction away from the observer. When many loops are joined together to form a coil, the magnetic flux lines surround the coil, as shown in Fig. 5-5. The field around a coil is much stronger than tile field of one loop of wire. The field around the coil is the same shape as the field around a bar magnet. A coil that has an iron or steel core inside it is called an electromagnet. A core increases the magnetic flux density of a coil. [8]

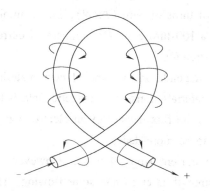
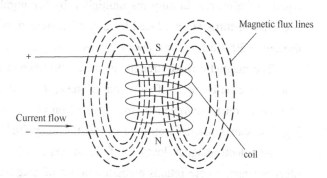

Fig. 5-4　Magnetic Filed around a Loop of Wire　　　Fig. 5-5　Magnetic Filed around a Coil of Wire

5.3.3　Electromagnets

Electromagnets are produced when current flows through a coil of wire. The north pole of a coil of wire is the end where the lines of force come out. The south pole is the end where the lines of force enter the coil. This is like the field of a bar magnet. To find the north pole of a coil, use the left-hand rule for polarity, as shown in Fig. 5-6. Grasp the coil with the left hand. Point the fingers in the direction of current flow through the coil. Tile thumb points to the north polarity of the coil.

When the polarity of the voltage source is reversed, the magnetic poles of tile coil also reverse. The poles of an electromagnet can be checked with a compass. The compass is placed near a pole of the electromagnet. If the north-seeking pole of the compass points to the coil, that side is the north pole.

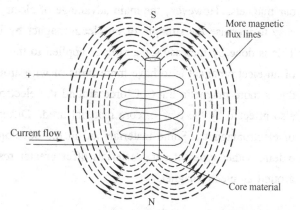

Fig. 5-6　Magnetic Filed around a Coil of Wire with a Core

Electromagnets have several turns of wire wound around a soft-iron core. An electrical power source is then connected to the ends of the turns of wire. When current flows through the wire, the magnetic polarities are produced at the ends of the soft iron core. [9] The three basic parts of an electromagnet are①an iron core, ②wire windings, and③an electrical power source. Electromagnetism is made possible by electric current flow, which produces a magnetic field. When electrical current flows through the coil, the properties of magnetic materials are developed.

The magnetic strength of an electromagnet depends on three factors: ① the amount of current passing through the coil, ② the number of turns of wire, and ③ the type of core material. [10] The number of magnetic lines of force is increased by increasing the current, by increasing the number of turns of wire, or by using a more desirable type of core material. The magnetic strength of electromagnets is determined by the ampere-turns of each coil. The number of ampere-turns is equal to the current in amperes multiplied by the number of turns of wire ($I \times N$). For example, 200 ampere-turns are produced by 2 A of current through a 100-turn coil. One ampere of current through a 200-turn coil produces the same magnetic field strength.

The magnetic field strength of an electromagnet also depends on the type of core material. Cores are usually made of soft iron or steel, since these materials transfer a magnetic field better than air or other nonmagnetic materials. Iron cores increase the flux density of an electromagnet. Fig. 5-6 shows that an iron core causes the magnetic flux to be more dense.

An electromagnet loses its field strength when the current stops flowing. However, an electromagnet's core retains a small amount of magnetic strength after current stops flowing. This is called residual magnetism, or "leftover" magnetism.[11] It can be reduced by using soft iron cores or increased by using hard-steel core material. Residual magnetism is very important in the operation of some types of electrical generators.

In many ways, electromagnetism is similar to magnetism produced by natural magnets such as bar magnets. However, the main advantage of electromagnetism is that it is easily controlled. It is easy to increase the strength of an electromagnet by increasing the current flow through tile coil. This is done by increasing the voltage applied to the coil. The second way to increase tile strength of an electromagnet is to have more turns of wire around tile core. A greater number of turns produces more magnetic lines of force around the electromagnet. The strength of an electromagnet is also affected by the type of core material used. Different alloys of iron are used to make the cores of electromagnets. Some materials aid in the development of magnetic lines of force to a greater extent. Other types of core materials offer greater resistance to the development of magnetic flux around an electromagnet.

5.4 Ohm's Law for Magnetic Circuits

A relationship similar to Ohm's law for electrical circuits exists in magnetic circuits. Magnetic circuits have magnetomotive force (MMF), magnetic flux (Φ), and reluctance (R). MMF is the force that causes a magnetic flux to be developed. Magnetic flux consists of the lines of force around a magnetic material.[12] Reluctance is the opposition to the development of a magnetic flux. These terms may be compared to voltage, current, and resistance in electrical circuits. When MMF increases, magnetic flux increases. Remember that in an electrical circuit, when voltage increases, current increases. When resistance in an electrical circuit increases, current decreases. When reluctance of a magnetic circuit increases, magnetic flux decreases.

5.5 Domain Theory of Magnetism

A theory of magnetism was presented in the nineteenth century by the German scientist Wilhelm Weber. Weber's theory of magnetism was called the molecular theory. It dealt with the alignment of molecules in magnetic materials. Weber felt that molecules were aligned in an orderly arrangement in magnetic materials. In nonmagnetic materials, he thought that molecules were arranged in a random pattern.

Weber's theory has now been modified somewhat to become the domain theory of magnetism. This theory deals with the alignment in materials of domains rather than molecules. [13] A domain is a group of atoms (about 10^{15} atoms). Each domain acts like a tiny magnet. The rotation of electrons around the nucleus of these atoms is important. Electrons have a negative charge. As they orbit around the nucleus of atoms, their electrical charge moves. This moving electrical field produces a magnetic field. The polarity of the magnetic field is determined by the direction of electron rotation.

The domains of magnetic materials are atoms grouped together. Their electrons are believed to spin in the same directions. This produces a magnetic field due to electrical charge movement. In nonmagnetic materials, half of the electrons spinin one direction and half spin in the opposite direction. Their charges cancel each other out, so no magnetic field is produced. Electron rotation in magnetic materials is in the same direction, which causes the domains to act like tiny magnets that align to produce a magnetic field.

5.6 Electricity Produced by Magnetism

A scientist named Michael Faraday discovered in the early 1830s that electricity is produced from magnetism. He found that if a magnet is placed inside a coil of wire, electrical current is produced when the magnet is moved.

Faraday's law is stated as follows: When a coil of wire moves across the lines of force of a magnetic field, electrons flow through the wire in one direction. When the coil of wire moves across the magnetic lines of force in the opposite direction, electrons flow through the wire in the opposite direction. [14] This is the principle of electrical power generation. Most of the electrical energy used today is produced by using magnetic energy.

Current flows in a conductor placed inside a magnetic field only when there is motion between the conductor and the magnetic field. If a conductor is stopped moving across the magnetic lines of force, currentwill stop flowing. This principle is called electromagnetic induction. [15] The operation of electrical generators depends on conductors moving across a magnetic field. The right-hand rule is used to determine the direction of electron flow. This rule for generators is stated as follows: Hold the thumb, forefinger, and middle finger of the right hand perpendicular to each other. Point the forefinger in the direction of the magnetic field from north to south. Point the thumb in the direction of the motion of the conductor. The middle finger will then point in the direction of electron current flow.

If a conductor or a group of conductors is moved through a strong magnetic field, induced current will flow and a voltage will be produced. Fig. 5-7 shows a loop of wire rotated through a magnetic field. The position of the loop inside the magnetic field determines the amount of induced current and voltage. The opposite sides of the loop move across the magnetic lines of force in opposite directions. This movement causes an equal amount of electrical current to flow in opposite directions through the two sides of the loop. Notice each position of the loop and the resulting output voltage in Fig. 5-7. The electrical current flows in one direction and then in the opposite direction with every complete revolution of the conductor. This method produces alternating current. One complete rotation is called a cycle. The number of cycles per second is known as the frequency. Most AC generators produce 50 cycles per second.[16]

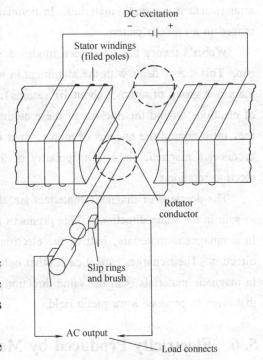

Fig. 5-7 Basic AC Generator

The ends of the conductor that move across the magnetic field of the generator shown in Fig. 5-7 are connected to slip rings and brushes. The slip rings are mounted on the same shaft as the conductor. Carbon brushes are used to make contact with the slip rings. The electrical current induced into the conductor flows through the slip rings to the brushes. When the conductor turns half a revolution, electrical current flows in one direction through the slip rings and the load.[17] During the next half revolution of the coil, the positions of the two sides of the conductor are opposite. The direction of the induced current is reversed. Current now flows through the load in the opposite direction.

The conductors that make up the rotor of a generator have many turns. The generated voltage is determined by these three factors:

1) The number of turns of wire used.
2) The strength of the magnetic field.
3) The speed of the prime mover used to rotate the machine.

Direct current can also be produced by electromagnetic induction. A simple DC generator has a split-ring commutator instead of two slip rings. The split rings resemble one full ring, except that they are separated by small openings. Induced electrical current still flows in opposite directions to each half of the split ring. However, current flows in the same direction in the load circuit due to the action of the split rings.

New Words and Expressions

1. magnetism   *n.* 磁，磁力，磁学
2. electromagnetism   *n.* 电磁，电磁学
3. magnet   *n.* 磁铁
4. magnetic flux   磁通量
5. clockwise   *adj.* 顺时针方向的
6. counterclockwise   *adj.* 逆时针方向的

7. compass   *n.* 指南针，罗盘
8. loop   *n.* 回路，线圈，环
9. coil   *n.* 线圈
10. turns   *n.* 匝数
11. winding   *n.* 绕组
12. flux density   磁通密度
13. magnetomotive force（MMF）   磁动势
14. molecular   *n.* 分子的，由分子组成的
15. align   *v.* 校准
16. generator   *n.* 发电机
17. magnetic field   磁场
18. induction   *n.* 感应，感应现象
19. slip ring   集电环，滑环
20. carbon   *n.* 碳
21. commutator   *n.* 换向器
22. reluctance   *n.* 磁阻

Notes

[1] Some metals in their natural state attract small pieces of iron. This property is called magnetism. Materials that have this ability are called natural magnets.

一些金属在天然状态下会吸引小块铁片，这种属性称为磁性。具有这种性能的材料称为天然磁铁。

[2] Electromagnetism is magnetism that is brought about due to electrical current flow. There are many electrical machines that operate because of electromagnetism.

电磁现象是由于电流流通所产生的磁力现象。有许多电机依据电磁学原理工作。

[3] The field is made up of lines of force, or magnetic flux. These magnetic flux lines are invisible. They never cross one another, but they always form individual closed loops around a magnetic material. They have a definite direction from the north to the south pole along the outside of a magnet.

场是由磁势线或磁力线组成的。磁力线是不可见的，彼此互不相交，但是它们总是在磁性材料周围形成各自的闭合回路。沿着磁铁的外围从北极到南极的方向定义为磁力线的方向。

[4] When magnetic flux lines are close together, the magnetic field is strong. When magnetic flux lines are farther apart, the field is weaker.

磁力线间隔近的地方，磁场强；磁力线间隔远的地方，磁场较弱。

[5] It is possible to magnetize pieces of metal temporarily by using a bar magnet. If a magnet is passed over the top of a piece of iron several times in the same direction, the soft iron becomes magnetized and stays magnetized for a short period of time.

通过使用一块条形磁铁使金属片暂时磁化是可能的。如果磁铁从一个铁块的顶端按一个方向通过几次，这块软磁性铁会被磁化，并且在短期内会保持磁性。

[6] A magnetic material can lose some of its magnetism if it is jarred or heated. People must be careful when handling equipment that contains permanent magnets.

磁性材料在剧烈振动或加热的情况下会损失部分磁性。因此人们在处理包含永久磁铁的控制设备时必须保持仔细。

[7] Current-carrying conductors produce a magnetic field. A compass is used to show that the magnetic flux lines are circular in shape. The conductor is in the center of the circular shape. The direction of the current flow and the magnetic flux lines can be shown by using the left-hand

rule of magnetic flux.

电流在导体中流通产生磁场。指南针可以用来表示环形磁力线的方向。导体位于环形磁力线的中心。电流流通的方向和磁力线的方向可以通过左手螺旋定则来表示。

[8] The field around a coil is much stronger than tile field of one loop of wire. The field around the coil is the same shape as the field around a bar magnet. A coil that has an iron or steel core inside it is called an electromagnet. A core increases the magnetic flux density of a coil.

线圈周围的磁场要比一个导线回路周围的磁场强得多。线圈周围磁场的外形与条形磁铁周围磁场的外形相同。带有铁心或钢心的线圈称为电磁铁。铁心使线圈的磁通密度增加。

[9] Electromagnets have several turns of wire wound around a soft-iron core. An electrical power source is then connected to the ends of the turns of wire. When current flows through the wire, the magnetic polarities are produced at the ends of the soft iron core.

电磁铁是指在软磁性铁心的周围捆扎几匝导线。然后电源与几匝导线的两端连接。当导线中有电流流通时就会在软磁性铁心的两端产生磁极。

[10] The magnetic strength of an electromagnet depends on three factors: ① the amount of current passing through the coil, ② the number of turns of wire, and ③ the type of core material.

电磁铁周围磁场的强度取决于三个因素：①流过线圈的电流值，②导线的匝数，③铁心材料的类型。

[11] An electromagnet loses its field strength when the current stops flowing. However, an electromagnet's core retains a small amount of magnetic strength after current stops flowing. This is called residual magnetism, or "leftover" magnetism.

当电流停止流通时，电磁铁就会失去场强。然而，电流停止流通后电磁铁的铁心仍会保留少量的磁场强度，这被称为残磁或剩磁。

[12] A relationship similar to Ohm's law for electrical circuits exists in magnetic circuits. Magnetic circuits have magnetomotive force (MMF), magnetic flux (Φ), and reluctance (R). MMF is the force that causes a magnetic flux to be developed. Magnetic flux consists of the lines of force around a magnetic material.

在磁路中存在与电路关系相似的欧姆定律。磁路有磁动势、磁通和磁阻。磁动势是导致磁通产生的力。磁通是由磁性材料周围的磁力线所组成的。

[13] Weber's theory has now been modified somewhat to become the domain theory of magnetism. This theory deals with the alignment in materials of domains rather than molecules.

韦伯理论已经做了某些修正而成为磁畴理论，这个理论涉及在材料中磁畴的排列而不是分子的排列。

[14] When a coil of wire moves across the lines of force of a magnetic field, electrons flow through the wire in one direction. When the coil of wire moves across the magnetic lines of force in the opposite direction, electrons flow through the wire in the opposite direction.

当线圈运动切割磁场的磁力线时，电子将按一个方向流动通过导线。当线圈切割磁力线的运动方向相反时，电子将按相反的方向流动通过导线。

[15] Current flows in a conductor placed inside a magnetic field only when there is motion between the conductor and the magnetic field. If a conductor is stopped moving across the magnet-

ic lines of force, current will stop flowing. This principle is called electromagnetic induction.

放在磁场中的导体,只有导体与磁场间有运动时导体中才会产生电流。如果导体停止切割磁力线的运动,导体中的电流将停止流动,这个原理称为电磁感应现象。

[16] The electrical current flows in one direction and then in the opposite direction with every complete revolution of the conductor. This method produces alternating current. One complete rotation is called a cycle. The number of cycles per second is known as the frequency. Most AC generators produce 50 cycles per second.

导体每旋转一圈,电流先按一个方向流动,然后按相反的方向流动。交流就是按这种方法产生。旋转一整圈称为一个周期。每秒的周期数就是众所周知的频率。大部分交流发电机发出的频率为每秒 50 周。

[17] The slip rings are mounted on the same shaft as the conductor. Carbon brushes are used to make contact with the slip rings. The electrical current induced into the conductor flows through the slip rings to the brushes. When the conductor turns half a revolution, electrical current flows in one direction through the slip rings and the load.

滑环与导体被安装在同一轴上,使用碳刷与滑环做接触连接。导体中感应的电流由滑环流向碳刷。导体旋转的每半个周期,电流按一个方向流过滑环和负载。

PART 2 CONTROL THEORY AND TECHNOLOGY

Chapter 6 Knowledge of Control Theory

6.1 What Is Control?

The term *control* has many meanings and often varies between communities. In this chapter, we define control to be the use of algorithms and feedback in engineered systems. Thus, control includes such examples as feedback loops in electronic amplifiers, setpoint controllers in chemical and materials processing, "fly-by-wire" systems on aircraft and even router protocols that control traffic flow on the Internet. Emerging applications include high-confidence software systems, autonomous vehicles and robots, real-time resource management systems and biologically engineered systems. At its core, control is an information science and includes the use of information in both analog and digital representations.[1]

A modern controller senses the operation of a system, compares it against the desired behavior, computes corrective actions based on a model of the system's response to external inputs and actuates the system to effect the desired change. This basic feedback loop of sensing, computation and actuation is the central concept in control.[2] The key issues in designing control logic are ensuring that the dynamics of the closed loop system are stable (bounded disturbances give bounded errors) and that they have additional desired behavior (good disturbance attenuation, fast responsiveness to changes in operating point, etc.). These properties are established using a variety of modeling and analysis techniques that capture the essential dynamics of the system and permit the exploration of possible behaviors in the presence of uncertainty, noise and component failure.

A typical example of a control system is shown in Fig. 6-1. The basic elements of sensing, computation and actuation are clearly seen. In modern control systems, computation is typically implemented on a digital computer, requiring the use of analog-to-digital (A/D) and digital-to-analog (D/A) converters.[3] Uncertainty enters the system through noise in sensing and actuation subsystems, external disturbances that affect the underlying system operation and uncertain dynamics in the system (parameter errors, unmodeled effects, etc). The algorithm that computes the control action as a function of the sensor values is often called a control law. The system can be influenced externally by an operator who introduces command signals to the system.

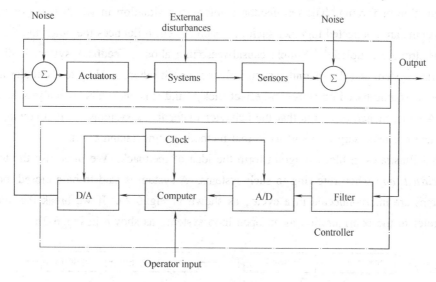

Fig. 6-1 Components of a Computer-Controlled System

Control engineering relies on and shares tools from physics (dynamics and modeling), computer science (information and software) and operations research (optimization, probability theory and game theory), but it is also different from these subjects in both insights and approach.

Perhaps the strongest area of overlap between control and other disciplines is in the modeling of physical systems, which is common across all areas of engineering and science. One of the fundamental differences between control-oriented modeling and modeling in other disciplines is the way in which interactions between subsystems are represented. [4] Control relies on a type of input/output modeling that allows many new insights into the behavior of systems, such as disturbance attenuation and stable interconnection. Model reduction, where a simpler (lower-fidelity) description of the dynamics is derived from a high-fidelity model, is also naturally described in an input/output framework. Perhaps most importantly, modeling in a control context allows the design of robust interconnections between subsystems, a feature that is crucial in the operation of all large engineered systems.

Control is also closely associated with computer science since virtually all modern control algorithms for engineering systems are implemented in software. However, control algorithms and software can be very different from traditional computer software because of the central role of the dynamics of the system and the real-time nature of the implementation.

6.2 Feedback

Feedback is a central feature of life. The process of feedback governs how we grow, respond to stress and challenge, and regulate factors. In this chapter we provide an introduction to the basic concept of feedback and the related engineering discipline of control.

A dynamical system is a system whose behavior changes over time, often in response to external stimulation or forcing. The termfeedback refers to a situation in which two (or more) dynamical systems are connected together such that each system influences the other and their dynamics are thus strongly coupled.[5] Simple causal reasoning about a feedback system is difficult because the first system and the second system influences each other, leading to a circular argument. This makes reasoning based on cause and effect tricky, and it is necessary to analyze the system as a whole. A consequence of this is that the behavior of feedback systems is often counterintuitive, and it is therefore necessary to resort to formal methods to understand them.

Fig. 6-2 illustrates in block diagram form the idea of feedback. We often use the terms *open loop* and *closed loop* when referring to such systems. A system is said to be a closed loop system if the systems are interconnected in a cycle, as shown in Fig. 6-2a. If we break the interconnection, we refer to the configuration as an open loop system, as shown in Fig. 6-2b.

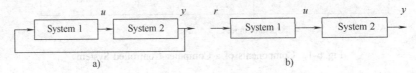

Fig. 6-2　Open and Closed Loop Systems
a) Closed Loop　b) Open Loop

An early engineering example of a feedback system is a centrifugal governor, in which the shaft of a steam engine is connected to a flyball mechanism that is itself connected to the throttle of the steam engine. The system is designed so that as the speed of the engine increases (perhaps because of a lessening of the load on the engine), the flyballs spread apart and a linkage causes the throttle on the steam engine to be closed. This in turn slows down the engine, which causes the flyballs to come back together. We can model this system as a closed loop system by taking System 1 as the steam engine and System 2 as the governor. When properly designed, the flyball governor maintains a constant speed of the engine, roughly independent of the loading conditions. The centrifugal governor was an enabler of the successful Watt steam engine, which fueled the industrial revolution.

Feedback has many interesting properties that can be exploited in designing systems. As in the case of the flyball governor, feedback can make a system resilient toward external influences. It can also be used to create linear behavior out of nonlinear components, a common approach in electronics. More generally, feedback allows a system to be insensitive both to external disturbances and to variations in its individual elements.[6]

Feedback has potential disadvantages as well. It can create dynamic instabilities in a system, causing oscillations or even runaway behavior.[7] Another drawback, especially in engineering systems, is that feedback can introduce unwanted sensor noise into the system, requiring careful filtering of signals. It is for these reasons that a substantial portion of the study of feedback systems is devoted to developing an understanding of dynamics and a mastery of techniques in dynamical systems.

Feedback systems are ubiquitous in both natural and engineered systems. Control systems maintain the environment, lighting and power in our buildings and factories; they regulate the operation of our cars, consumer electronics and manufacturing processes; they enable our transportation and communications systems; and they are critical elements in our military and space systems. For the most part they are hidden from view, buried within the code of embedded microprocessors, executing their functions accurately and reliably. Feedback has also made it possible to increase dramatically the precision of instruments such as Atomic Force Microscopes (AFMs) and telescopes.

Feedback is a powerful idea which, as we have seen, is used extensively in natural and technological systems. The principle of feedback is simple: base correcting actions on the difference between desired and actual performance. In engineering, feedback has been rediscovered and patented many times in many different contexts. The use of feedback has often resulted in vast improvements in system capability, and these improvements have sometimes been revolutionary, as discussed above. [8] The reason for this is that feedback has some truly remarkable properties. In this section we will discuss some of the properties of feedback that can be understood intuitively.

6.3 PID Control

The reason why on-off control often gives rise to oscillations is that the system overreacts since a small change in the error makes the actuated variable change over the full range. This effect is avoided in proportional control, where the characteristic of the controller is proportional to the control error for small errors. This can be achieved with the control law

$$u = \begin{cases} u_{max} & if \quad e \geqslant e_{max} \\ k_p e & if \quad e_{min} < e < e_{max} \\ u_{max} & if \quad e \leqslant e_{min} \end{cases} \quad (6\text{-}1)$$

Where k_p is the controller gain, $e_{min} = u_{min}/k_p$ and $e_{max} = u_{max}/k_p$, the interval (e_{min}, e_{max}) is called the *proportional band* because the behavior of the controller is linear when the error is in this interval:

$$u(t) = k_p(r - y) = k_p e \quad (6\text{-}2)$$

While a vast improvement over on-off control, proportional control has the drawback that the process variable often deviates from its reference value. In particular, if some level of controlsignal is required for the system to maintain a desired value, then we must have $e \neq 0$ in order to generate the requisite input.

This can be avoided by making the control action proportional to the integral of the error

$$u(t) = K_i \int_0^t e(\tau) d\tau \quad (6\text{-}3)$$

This control form is called integral control, and k_i is the integral gain. It can be shown through simple arguments that a controller with integral action has zero steady-state error. The catch is that there may not always be a steady state because the system may be oscillating.

An additional refinement is to provide the controller with an anticipative ability by using a prediction of the error. A simple prediction is given by the linear extrapolation

$$e(t+T_d) \approx e(t) + T_d \frac{de(t)}{dt} \tag{6-4}$$

Which predicts the error T_d time units ahead. Combining proportional, integral and derivative control, we obtain a controller that can be expressed mathematically as

$$u(t) = K_p e(t) + K_i \int_0^t e(\tau)d(\tau) + k_d \frac{de(t)}{dt} \tag{6-5}$$

The control action is thus a sum of three terms: the past as represented by the integral of the error, the present as represented by the proportional term and the future as represented by a linear extrapolation of the error (the derivative term). [9] This form of controller is called a Proportional-Integral-Derivative (PID) controller.

A PID controller is very useful and is capable of solving a wide range of control problems. More than 95% of all industrial control problems are solved by PID control, although many of these controllers are actually Proportional-Integral (PI) controllers because derivative action is often not included. [10] There are also more advanced controllers, which differ from PID controllers by using more sophisticated methods for prediction.

6.4 Adaptive Control

6.4.1 Introduction

For over forty years the field of adaptive control and adaptive systems has been an attractive and developing area for theorists, researchers, and engineers. Over 6000 publications have appeared during the history of adaptive systems and this number is certainly not definitive. However, the number of industry applications is still low because many manufacturers distrust nontraditional and sometimes rather complicated methods of control. The classic methods of control and regulation have often been preferred. These have been worked out in detail, tested, and in many cases reached the desired reliability and quality. On the other hand, it is necessary to realize that the vast majority of processes to be controlled are, in fact, neither linear nor stationary systems and change their characteristics over time or when the set point changes. [11] Such changes affect different processes in various ways and are not always significant. In other systems and processes, however, the changes may be significant enough to make the use of controllers with fixed parameters, particularly of a PID type, unacceptable or eventually impossible.

The first attempts to develop a new and higher quality type of controller, capable of adapting and modifying its behaviour as conditions change (due to stochastic disturbances), were made in the 1950s. This was in the construction of autopilot systems, in aeronautics, in the air force, and in the military. The concept of adaptation that in living organisms is characterized as a process of adaptation and learning was thus transferred to control, technical, and cybernetic systems.

From the beginning the development of adaptive systems was extremely heterogeneous and fruitful. [12] The results were dependent on the level of the theory used and the technical and computing equipment available. During the first attempts simple analogue techniques (known as MIT algorithms) were applied. Later algorithms became more complicated and the theory, more demanding. Many approaches were not suitable for real-time applications because of the performance of the available computers or were simply too sophisticated for analogue computers. That period brought results of a theoretical research value only. References show the wealth of methods and approaches dating from this pioneering age.

During the 1960s two main areas emerged from this diversity of approaches to dominate the field of adaptive systems for many years. The first were Model Reference Adaptive Systems (MRAS) in which the parameters of the controller modify themselves so that the feedback system has the required behaviour. The second were self-tuning controllers (STC), which due to use of the matrix inversion lemma were able to use measured data to identify the model (the controlled process) on-line. The linear feedback controller parameters adapt according to the values of the identified parameters of the process model. Naturally both directions had their own supporters. The next decade was characterized by growing attempts to use adaptive systems in real-world applications, increased use of modern computers, and by applying the latest information and methods available in the theory of control. Examples include:

1) The use of algebraic approach in control design;
2) The parameterization of controllers;
3) The use of rational fraction functions;
4) The digitalization of signals and models.

A survey of developments during this period is given in references. The 1980s saw further breakthroughs. As microprocessor became ever faster, cheaper and more compact, and analogue equipment began to fall into disuse, the level of digitalization increased and became attractive for real-time use. [13] Adaptation was also relevant to other areas such as filtration, signal prediction, image recognition, and others. Methods known as auto-tuning started to appear, in which adaptation only occurs in the first stage of control (in order to identify the right controller type, which then remains fixed). Conferences dealing specifically with adaptive systems were held and the number of monographs and special publications increased. In the early 1990s new discoveries were applied to adaptive methods, such as artificial intelligence, neuron networks and fuzzy techniques.

However, in the second half of the 1990s adaptive systems still showed great unused potential in mass applications even though many well-known companies deployed adaptive principles for auto-tuning and occasionally even for on-line control. There were still opportunities for improvements, for streamlining in the areas of theory and application, and for increasing reliability and robustness. It has been accepted that there are processes that can only be controlled with automatic adaptive controllers but that are still controlled manually. In many real world processes this high quality control must be ensured and as the process alters, this leads back to adaptation. It is reasonable to assume that even where a nonadaptive controller is sufficient, an adaptive controller can achieve improvements in the quality of control. An example of this is the use of an adaptive control decreased fuel consumption significantly.

6.4.2 Formulation of Adaptive Control Problem

Originally, adaptation was displayed only by plants and animals, where it is seen in its most varied forms. It is a characteristic of living organisms that they adapt their behaviour to their environment even where this is harsh. [14]

Each adaptation involves a certain loss for the organism, whether it is material, energy, or information. After repeated adaptations to the same changes, plants and animals manage to keep such losses to a minimum. Repeated adaptation is, in fact, an accumulation of experiences that the organism can evaluate to minimize the losses involved in adaptation. We call this learning.

Alongside such systems found in nature there are also technical systems capable of adaptation. These vary greatly in nature, and a wide range of mathematical tools are used to describe them. It is therefore impossible to find a single mathematical process to define all adaptive systems. For the purposes of our definition of adaptive systems we will limit ourselves to cybernetic systems which meet the following assumptions:

1) Their state or structure may change;
2) We may influence the state or output of the system.

One possible generalized definition of an adaptive system is as follows:

The adaptive system has three inputs and one output (Fig. 6-3). The environment acting on the adaptive system is composed of two elements: the reference variable w and disturbance v. The reference variable is created by the user but, as a rule, the disturbance cannot be measured. The system receives information on the required behaviour Ω, the system output is the behaviour of the system (decided rule)

$$y = f(w,v,\Theta) \qquad (6-6)$$

Which assigns the single outputy to each behaviour occurring in environments w and v. A change in behaviour, i.e. a change in this functionality, is effected by changing parameters Θ. For each combination (w, v, Θ) we select parameter $\Theta *$ in place of Θ so as to minimize loss function g (for unit time or for a given time period)

$$g(\Omega,w,v,\Theta) - g_{\min}(\Omega,w,v,\Theta *) \qquad (6-7)$$

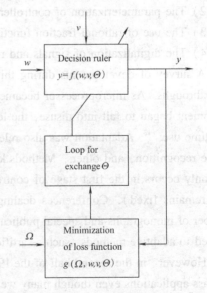

Fig. 6-3 Inner Structure of an Adaptive System

In this case adaptation is the process used to search for $\Theta *$ and continues until this parameter is found. A characteristic property of an adaptive system is the fact that the process of adaptation always occurs when there is a change in the environment w or v or a change in the required behaviour. If a change occurs after each time interval T_o adaptation will take place repeatedly at the start of each interval. If the adaptation then lasts for time t (after which loss g decreases) then the mean loss will be lower with a smaller ratio t/T_o. The inverse value of the mean loss is known as the adaptation effect.

We mention here the so-called learning system. The learning system can be seen as a system that remembers the optimal value of parameter $\Theta *$ on finishing the adaptation for the given m triplet (w_m, v_m, Ω_m) of sequence $\{(w_m, v_m, \Omega_m)\}$, for $k = 1, 2, \cdots, m, \cdots, \infty$, and uses it to create in its memory the following function

$$\Theta * = f(w, v, \Omega) \qquad (6-8)$$

On completing the learning process the decided rule for every behaviour, in environments w and v can be chosen directly by selecting the appropriate value for parameter $\Theta *$ from memory without adaptation.

We can conclude, therefore, that an adaptive system constantly repeats the adaptation process, even when the environment behaviour remains unchanged, and needs constant information on the required behaviour. A learning system evaluates repeated adaptations so as to remember any state previously encountered during adaptation and when this reoccurs in the environment, does not use Equation (6-7) to find the optimum but uses information already in its memory.

Adaptive and learning systems can be used to solve the following tasks:

1) Recursive identification—i. e. the creation of a mathematical description of the controlled process using self-adjusting models.

2) The control of systems about which we know too little before starting up to predefine the structure and parameters of the control algorithm, and also systems whose transfer characteristics change during control.

3) Recognition of subjects or situations (scenes) and their classification. Adaptive and learning systems are then components of so-called classifiers.

4) Manipulation of subjects—i. e. change of their spatial position. Adaptive and learning systems are then components of robots.

Further we will focus only on problems of adaptive control. Fig. 6-4 shows a general block diagram of an adaptive system. According to this diagram we can formulate the following definition:

An adaptive system measures particular features of the adjustable

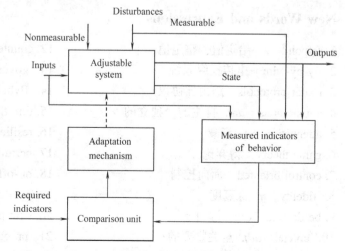

Fig. 6-4 General Block Diagram of Adaptive Control System

system behaviour using its inputs, states and outputs. By virtue of comparison of these measured features and sets of required features it modifies parameters and the structure of an adjustable loop or generates an auxiliary input so that the measured features track as closely as possible the required features.

This definition is fairly general and allows inclusion of most of the adaptive problems of technical cybernetics.[15] Features of the behaviour can take different forms in these problems. If the adaptive system is used for control, the behaviour feature could be, for example:

1) Pole and zeros assignment of a closed loop system;
2) The required overshoot of the step response of a closed loop system to reference and input disturbances;
3) The settling time;
4) The minimum value of various integral or summing criteria;
5) The amplitude and natural frequency of oscillations in nonlinear loops;
6) The frequency spectrum of a closed loop control system;
7) The required value of gain and phase margins, etc..

For the purposes of automatic control we can simplify the definition of an adaptive system still further:

Adaptive control systems adapt the parameters or structure of one part of the system (the controller) to changes in the parameters or structure in another part of the system (the controlled system) in such a way that the entire system maintains optimal behaviour according to the given criteria, independent of any changes that might have occurred.

Adaptation to changes in the parameters or structure of the system can basically be performed in three ways:

1) By making a suitable alteration to tile adjustable parameters of controller;
2) By altering the structure of the controller;
3) By generating a suitable auxiliary input signal (an adaptation by signal).

New Words and Expressions

1. setpoint   *n.* 给定值，整定值
2. fly-by-wire   电传操纵系统
3. router protocols   路由器协议
4. autonomous   *adj.* 自主的，独立的
5. attenuation   *n.* 衰减
6. game theory   博弈论
7. control-oriented   面向控制
8. fidelity   *n.* 保真度
9. be derived from   源自于
10. crucial   *adj.* 至关紧要的
11. causal reasoning   因果推理
12. counterintuitive   *adj.* 违反直觉的
13. governor   *n.* 调节器
14. flyball   *n.* 飞锤，离心球
15. throttle   *n.* 油门
16. resilient   *adj.* 有弹性的，有回力的
17. perturbation   *n.* 扰动，干扰
18. on-off control   开关式控制
19. over-react   过度反应
20. extrapolation   *n.* 归纳，推论
21. proportional-integral-derivative (PID) controller   比例-积分-微分（PID）控制器

22. stochastic disturbance　随机扰动
23. autopilot　*n.* 自动驾驶（仪）
24. aeronautics　*n.* 航空学
25. cybernetic　*n.* 控制论
26. heterogenous　*adj.* 不同种类的，多相的
27. diversity　*n.* 差异，多样性
28. fuzzy control　模糊控制
29. matrix inversion　矩阵求逆
30. lemma　*n.* 引理
31. rational fraction　有理分式
32. monograph　*n.* 专论
33. harsh　*adj.* 粗糙的，苛刻的
34. accumulation　*n.* 积累，累加值
35. learning system　学习系统，训练系统
36. recursive　*adj.* 递归的，循环的
37. identification　*n.* 辨识，识别
38. self-adjusting　自调节
39. criteria　*n.* 标准
40. frequency spectrum　频谱

Notes

［1］At its core, control is an information science and includes the use of information in both analog and digital representations.

其核心是，控制是一门信息科学，它既包含模拟形式又包含数字形式信息的应用。

［2］This basic feedback loop of sensing, computation and actuation is the central concept in control.

这个由检测、传感和执行机构构成的基本反馈回路是控制（理论）的核心概念。

［3］In modern control systems, computation is typically implemented on a digital computer, requiring the use of analog-to-digital (A/D) and digital-to-analog (D/A) converters.

在现代控制系统中，计算通常是通过数字计算机实现的，这就需要使用模/数转换器（A/D 转换器）和数/模转换器（D/A 转换器）。

［4］One of the fundamental differences between control-oriented modeling and modeling in other disciplines is the way in which interactions between subsystems are represented.

面向控制的建模与其他学科建模的一个根本区别是：子系统之间的相互作用的表示方式。

［5］The term *feedback* refers to a situation in which two (or more) dynamical systems are connected together such that each system influences the other and their dynamics are thus strongly coupled.

"反馈"一词指的是两个（或多个）动态系统之间相互连接时，每一个系统都对其他系统产生影响，它们之间的动态性能是强耦合的。

［6］More generally, feedback allows a system to be insensitive both to external disturbances and to variations in its individual elements.

总体说来，反馈使得系统不仅对于外部扰动不敏感，对于自身的变化也不敏感。

［7］Feedback has potential disadvantages as well. It can create dynamic instabilities in a system, causing oscillations or even runaway behavior.

反馈也有潜在的缺陷，即在系统中可能导致动态不稳定，引起振荡甚至失去控制。

［8］The use of feedback has often resulted in vast improvements in system capability, and these improvements have sometimes been revolutionary, as discussed above.

反馈的使用通常使（系统）性能得到显著提高，这种提高有时候甚至是革命性的，正如前文所述。

[9] The control action is thus a sum of three terms: the past as represented by the integral of the error, the present as represented by the proportional term and the future as represented by a linear extrapolation of the error (the derivative term).

因此，控制作用包括三个方面：代表过去的积分误差，代表现在的比例项以及代表未来的误差线性外推（微分项）

[10] More than 95% of all industrial control problems are solved by PID control, although many of these controllers are actually Proportional-Integral (PI) controllers because derivative action is often not included.

超过95%的工业控制问题是通过PID控制器解决的，虽然经常由于不包含微分控制，实际采用的是PI控制器。

[11] On the other hand, it is necessary to realize that the vast majority of processes to be controlled are, in fact, neither linear nor stationary systems and change their characteristics over time or when the set point changes.

另一方面，需要认识到绝大多数的控制过程事实上既非线性，也非静态系统，它们的特性随着时间的变化或给定值的变化而变化。

[12] From the beginning the development of adaptive systems was extremely heterogeneous and fruitful.

从一开始，自适应系统的发展就是非常异类并且富有成果的。

[13] As microprocessor became ever faster, cheaper and more compact, and analogue equipment began to fall into disuse, the level of digitalization increased and became attractive for real-time use.

随着微处理器变得更快、更廉价、更紧凑，模拟设备开始被弃用，数字化水平不断提高，并在实时处理中变得更有吸引力。

[14] It is a characteristic of living organisms that they adapt their behaviour to their environment even where this is harsh.

这是一种生物组织的特征，即它们能使自己的行为适应环境，即使是很恶劣的环境。

[15] This definition is fairly general and allows inclusion of most of the adaptive problems of technical cybernetics.

这个定义具有相当的普遍性，包含了大多数的技术控制论的自适应问题。

Chapter 7　Motor Drives and Controls

7.1　DC Motor Drives

The thyristor DC drive remains an important speed-controlled industrial drive, especially where the higher maintenance cost associated with the DC motor brushes (c. f. induction motor) is tolerable. The controlled (thyristor) rectifier provides a low-impedance adjustable 'DC' voltage for the motor armature, thereby providing speed control.

Until the 1960s, the only really satisfactory way of obtaining the variable-voltage DC supply needed for speed control of an industrial DC motor was to generate it with a DC generator. The generator was driven at fixed speed by an induction motor, and the field of the generator was varied in order to vary the generated voltage.[1] The motor/generator (MG) set could be sited remote from the DC motor, and multi-drive sites (e. g. steelworks) would have large rooms full of MG sets, one for each variable-speed motor on the plant. Three machines (all of the same power rating) were required for each of these Ward Leonard drives, which was good business for the motor manufacturer. For a brief period in the 1950s they were superseded by grid-controlled mercury arc rectifiers, but these were soon replaced by thyristor converters which offered cheaper first cost, higher efficiency (typically over 95%), smaller size, reduced maintenance, and faster response to changes in set speed. The disadvantages of rectified supplies are that the waveforms are not pure DC, that the overload capacity of the converter is very limited, and that a single converter is not capable of regeneration.

Though no longer pre-eminent, study of the DC drive is valuable for several reasons:

1) The structure and operation of the DC drive are reflected in almost all other drives, and lessons learned from the study of the DC drive therefore have close parallels to other types.

2) The DC drive tends to remain the yardstick by which other drives are judged.

3) Under constant-flux conditions the behaviour is governed by a relatively simple set of linear equations, so predicting both steady-state and transient behaviour is not difficult. When we turn to the successors of the DC drive, notably the induction motor drive, we will find that things are much more complex, and that in order to overcome the poor transient behaviour, the strategies adopted are based on emulating the DC drive.

7.1.1　Thyristor DC Drives-General

For motors up to a few kilowatts the armature converter can be supplied from either single-phase or three-phase mains, but for larger motors three-phase is always used. A separate thyristoror diode rectifier is used to supply the field of the motor: the power is much less than the armature power, so the supply is often single-phase, as shown in Fig. 7-1.

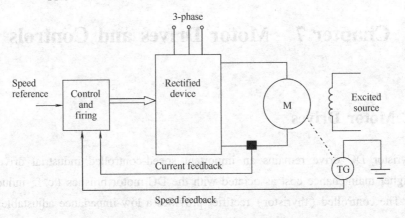

Fig. 7-1 Schematic Diagram of Speed-Controlled DC Motor Drive

The arrangement shown in Fig. 7-1 is typical of the majority of DC drives and provides for closed-loop speed control. The function of the two control loops will be explored later.

The main power circuit consists of a six-thyristor bridge circuit, which rectifies the incoming AC supply to produce a DC supply to the motor armature. The assembly of thyristors, mounted on a heat sink, is usually referred to as the stack. By altering the firing angle of the thyristors the mean value of the rectified voltage can be varied, thereby allowing the motor speed to be controlled.[2]

We know that the controlled rectifier produces a crude form of DC with a pronounced ripple in the output voltage. This ripple component gives rise to pulsating currents and fluxes in the motor, and in order to avoid excessive eddy-current losses and commutation problems, the poles and frame should be of laminated construction. It is accepted practice for motors supplied for use with thyristor drives to have laminated construction, but older motors often have solid poles and/or frames, and these will not always work satisfactorily with a rectifier supply. It is also the norm for drive motors to be supplied with an attached blower motor as standard. This provides continuous through ventilation and allows the motor to operate continuously at full torque even down to the lowest speeds without overheating.

Low power control circuits are used to monitor the principal variables of interest (usually motor current and speed), and to generate appropriate firing pulses so that the motor maintains constant speed despite variations in the load. The speed reference (Fig. 7-1) is typically an analogue voltage varying from 0 to 10V, and obtained from a manual speed-setting potentiometer or from elsewhere in the plant.

The combination of power, control, and protective circuits constitutes the converter. Standard modular converters are available as off-the-shelf items in sizes from 0.5kW up to several hundred kW, while larger drives will be tailored to individual requirements. Individual converters may be mounted in enclosures with isolators, fuses etc., or groups of converters may be mounted together to form a multi-motor drive.

1. Motor Operation with Converter Supply

We have known the basic operation of the rectifying bridge, now we turn to the matter of how the DC motor behaves when supplied with DC from a controlled rectifier.

By no stretch of imagination could the waveforms of armature voltage be thought of as good DC, and it would not be unreasonable to question the wisdom of feeding such an unpleasant looking waveform to a DC motor. In fact it turns out that the motor works almost as well as it would if fed with pure DC, for two main reasons. Firstly, the armature inductance of the motor causes the waveform of armature current to be much smoother than the waveform of armature voltage, which in turn means that the torque ripple is much less than might have been feared. And secondly, the inertia of the armature is sufficiently large for the speed to remain almost steady despite the torque ripple. It is indeed fortunate that such a simple arrangement works so well, because any attempt to smooth-out the voltage waveform would prove to be prohibitively expensive in the power ranges of interest.

2. Motor Current Waveforms

For the sake of simplicity we will look at operation from a three-phase (3-pulse) converter. The voltage applied to the motor armature is typically as shown in Fig. 7-2: it consists of rectified chunks of the incoming mains voltage, the precise shape and average value depending on the firing angle.

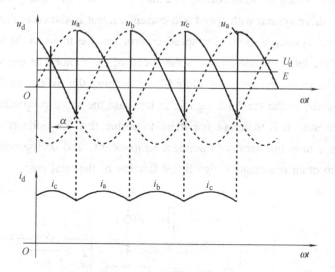

Fig. 7-2 Waveforms for Armature Voltage and Armature Current

The voltage waveform can be considered to consist of a mean DC level (U_d), and a superimposed pulsating or ripple component.[3] The mean voltage U_d can be altered by varying the firing angle, which also incidentally alters the ripple.

The ripple voltage causes a ripple current to flow in the armature, but because of the armature inductance, the amplitude of the ripple current is small.[4] In other words, the armature presents a high impedance to AC voltages. This smoothing effect of the armature inductance is shown in Fig. 7-2, from which it can be seen that the current ripple is relatively small in comparison with the corresponding voltage ripple. The average value of the ripple current is of course zero, so it has no effect on the average torque of the motor. There is nevertheless a variation in torque every half-cycle of the mains, but because it is of small amplitude and high frequency the variation in speed (and hence back EMF, E) will not usually be noticeable.

The current at the end of each pulse is the same as at the beginning, so it follows that the average voltage across the armature inductance (L) is zero. We can therefore equate the average applied voltage to the sum of the back EMF (assumed pure DC because we are ignoring speed fluctuation) and the average voltage across the armature resistance, to yield

$$U_d = E + I_d R \tag{7-1}$$

which is exactly the same as for operation from a pure DC supply. This is very important, as it underlines the fact that we can control the mean motor voltage, and hence the speed, simply by varying the converter delay angle.

7.1.2 Control Arrangements for DC Drives

The most common arrangement, which is used with only minor variations from small drives of say 0.5kW up to the largest industrial drives of several megawatts, is the so-called two-loop control. This has an inner feedback loop to control the current (and hence torque) and an outer loop to control speed. When position control is called for, a further outer position loop is added.

A standard DC drive system with speed and current control is shown in Fig. 7-3. The primary purpose of the control system is to provide speed control, so the input to the system is the speed reference signal on the left, and the output is the speed of the motor (as measured by the tachogenerator TG) on the right. As with any closed-loop system, the overall performance is heavily dependent on the quality of the feedback signal, in this case the speed-proportional voltage provided by the tachogenerator. It is therefore important to ensure that the tacho is of high quality (so that its output voltage does not vary with ambient temperature, and is ripple-free) and as a result the cost of the tacho often represents a significant fraction of the total cost.

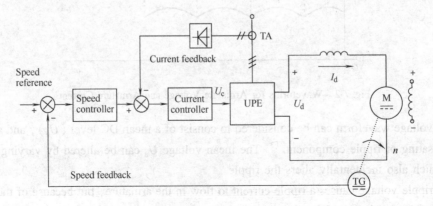

Fig. 7-3 Schematic Diagram of Analogue Controlled-Speed Drive with Current and Speed Feedback Control Loops

We will take an overview of how the scheme operates examine the function of the two loops in more detail.

To get an idea of the operation of the system we will consider what will happen if, with the motor running light at a set speed, the speed reference signal is suddenly increased. Because the set (reference) speed is now greater than the actual speed there will be a speed error signal, represented by the output of the left-hand summing junction in Fig. 7-3. A speed error indicates that acceleration is required, which in turn means torque, i. e. more current. The speed error is amplified by the speed controller (which is more accurately described as a speed-error amplifier) and the output serves as the reference or input signal to the inner control system. [5] The inner feedback loop is a current-control loop, so when the current reference increases, so does the motor armature current, thereby providing extra torque and initiating acceleration. As the speed rises the speed error reduces, and the current and torque therefore reduce to obtain a smooth approach to the target speed.

7.1.3 Chopper-Fed DC Motor Drives

If the source of supply is DC, a chopper-type converter is usually employed. The principal difference between the thyristor-controlled rectifier and the chopper is that in the former the motor current always flows through the supply, whereas in the latter, the motor current only flows from the supply terminals for part of each cycle.

A single-switch chopper using a transistor, MOSFET or IGBT can only supply positive voltage and current to a DC motor, and is therefore restricted to quadrant 1 motoring operation. [6] When regenerative and/or rapid speed reversal is called for, more complex circuitry is required, involving two or more power switches, and consequently leading to increased cost. Many different circuits are used and it is not possible to go into detail here.

1. Performance of Chopper-Fed DC Motor Drives

We saw earlier that the DC motor performed almost as well when fed from a phase-controlled rectifier as it does when supplied with pure DC. The chopper-fed motor is, if anything, rather better than the phase-controlled, because the armature current ripple can be less if a high chopping frequency is used. Typical waveforms of armature voltage and current are shown in Fig. 7-4b: these are drawn with the assumption that the switch is ideal. A chopping frequency of around 100Hz, as shown in Fig. 7-4, is typical of medium and large chopper drives, while small drives often use a much higher chopping frequency, and thus have lower ripple current. As usual, we have assumed that the speed remains constant despite the slightly pulsating torque, and that the armature current is continuous.

The shape of the armature voltage waveform reminds us that when the transistor is switched on, the battery voltage V is applied directly to the armature, and for the remainder of the cycle the transistor is turned off and the current freewheels through the diode. [7] When the current is freewheeling through the diode, the armature voltage is clamped at (almost) zero.

The speed of the motor is determined by the average armature voltage V_{DC}, which in turn depends on the proportion of the total cycle time (T) for which the transistor is on. If the on and off times are defined as defined as $T_{on} = kT$ and $T_{off} = (1-k)T$, where $0 < k < 1$, average voltage is simply given by $V_{DC} = kV$, from which we see that speed control is effected via the on time ratio, k.

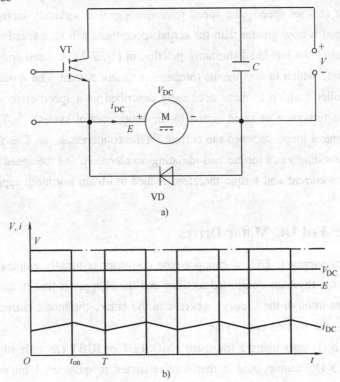

Fig. 7-4 Chopper-Fed DC Motor (t = 1oms)
a) A Simple Chopper Circuit b) The Typical Armature Voltage and Current Waveforms

Turning now to the current waveforms shown in Fig. 7-4b, the upper waveform corresponds to full load, i.e. the average current (I_{DC}) produces the full rated torque of the motor. If now the load torque on the motor shaft is reduced to half rated torque, and assuming that the resistance is negligible, the steady-state speed will remain the same but the new mean steady-state current will be halved, as shown by the lower curve. We note however that although, as expected, the mean current is determined by the load, the ripple current is unchanged, and this is explained below.

If we ignore resistance, the equation during 'on period' is

$$V = E + L\frac{di}{dt} \tag{7-2}$$

Since V is greater than E, the gradient of the current (di/dt) is positive, as can be seen in Fig. 7-4b. During this on period the battery is supplying power to the motor. Some of the energy is converted to mechanical output power, but some is also stored in the magnetic field associated with the inductance, the latter is given by $(1/2)Li^2$. We note that during the off time the gradient of the current is negative (as shown in Fig. 7-4b) and it is determined by the motional EMF E. During this period, the motor is producing mechanical output power which is supplied from the energy stored in the inductance; not surprisingly the current falls as the energy previously stored in the 'on' period is now given up.

During the off period, the equation governing the current is

$$0 = E + L \frac{di}{dt} \tag{7-3}$$

We note that the rise and fall of the current (i.e. the current ripple) is inversely proportional to the inductance, but is independent of the mean DC current, i.e. the ripple does not depend on the load.

2. Torque-Speed Characteristics and Control Arrangements

Under open-loop conditions (i.e. where the mark-space ratio of the chopper is fixed at a particular value) the behavior of the chopper-fed motor is similar to the converter-fed motor discussed earlier. When the armature current is continuous the speed falls only slightly with load, because the mean armature voltage remains constant. But when the armature current is discontinuous (which is most likely at high speeds and light load) the speed falls off rapidly when the load increases, because the mean armature voltage falls as the load increases. Discontinuous current can be avoided by adding an inductor in series with the armature, or by raising the chopping frequency, but when closed-loop speed control is employed, the undesirable effects of discontinuous current are masked by the control loop.

The control philosophy and arrangements for a chopper-fed motor are the same as for the converter-fed motor, with the obvious exception that the mark-space ratio of the chopper is used to vary the output voltage, rather than the firing angle. [8]

7.2 Inverter-Fed Induction Motor Drives

7.2.1 Introduction

As we know that the induction motor can only run efficiently at low slips, i.e. close to the synchronous speed of the rotating field. The best method of speed control must therefore provide for continuous smooth variation of the synchronous speed, which in turn calls for variation of the supply frequency. This is achieved using an inverter to supply the motor. A complete speed control scheme which includes tacho (speed) feedback is shown in block diagram form in Fig. 7-5.

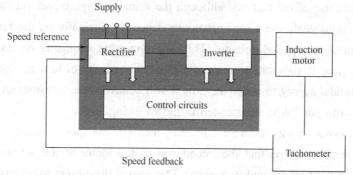

Fig. 7-5 General Arrangement of Inverter-Fed Variable-Frequency Induction Motor Speed-Controlled Drive

We should recall that the function of the converter is to draw power from the fixed-frequency constant-voltage mains, and convert it to variable frequency, variable voltage for driving the induction motor. Both the rectifier and the inverter employ switching strategies, so the power conversions are accomplished efficiently and the converter can be compact.

Variable frequency inverter-fed induction motor drives are used in ratings up to hundreds of kilowatts. Standard 50Hz or 60Hz motors are often used (though as we will see later this limits performance), and the inverter output frequency typically covers the range from around 5-10Hz up to perhaps 120Hz. This is sufficient to give at least a 10 : 1 speed range with a top speed of twice the normal (mains frequency) operating speed. The majority of inverters are 3-phase input and 3-phase output, but single-phase input versions are available up to about 5kW and some very small inverters (usually less than 1kW) are intended for use with single-phase motors.

A fundamental aspect of any converter, which is often overlooked, is the instantaneous energy balance. In principle, for any balanced 3-phase load, the total load power remains constant from instant to instant, so if it was possible to build an ideal 3-phase input, 3-phase output converter, there would be no need for the converter to include any energy storage elements. In practice, all converters require some energy storage, but these are relatively small when the input is 3-phase because the energy balance is good. However, as mentioned above, many small and medium power converters are supplied from single-phase mains. In this case, the instantaneous input power is zero at least twice per cycle of the mains. If the motor is 3-phase, it is obviously necessary to store sufficient energy in the converter to supply the motor during the brief intervals when the load power is greater than the input power. This explains why the most bulky components in many small and medium power inverters are electrolytic capacitors.

The majority of inverters used in motor drives are voltage source inverters (VSI), in which the output voltage to the motor is controlled to suit the operating conditions of the motor. Current source inverters (CSI) are still used, particularly for large applications, but will not be discussed here.

1. Inverter Waveforms

When we looked at the converter-fed DC motor, we saw that the behaviour was governed primarily by the mean DC voltage, and that for most purposes we could safely ignore the ripple components. A similar approximation is useful when looking at how the inverter-fed induction motor performs. We make use of the fact that although the actual voltage waveform supplied by the inverter will not be sinusoidal, the motor behaviour depends principally on the fundamental (sinusoidal) component of the applied voltage. This is a somewhat surprising but extremely welcome simplification, because it allows us to make use of our knowledge of how the induction motor behaves with a sinusoidal supply to anticipate how it will behave when fed from an inverter. A typical voltage waveforms for PWM inverter-fed is shown in Fig. 7-6.

In essence, the reason why the harmonic components of the applied voltage are much less significant than the fundamental is that the impedance of the motor at the harmonic frequencies is much higher than at the fundamental frequency. This causes the current to be much more sinusoidal than the voltage, and this in turn means that we can expect a sinusoidal traveling field to be set up.

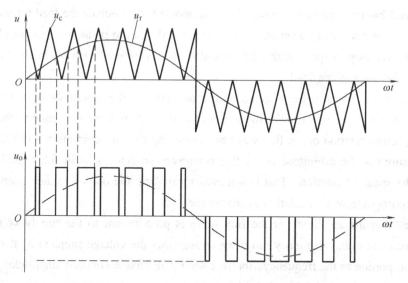

Fig. 7-6 Typical Voltage Waveforms for PWM Inverter-Fed

It would be wrong to pretend that the harmonic components have no effects, of course. They can create unpleasant acoustic noise, and always give rise to additional iron and copper losses.[9] As a result it is common for a standard motor to have to be derated for use on an inverter supply.

As with the DC drive the inverter-fed induction motor drive will draw non-sinusoidal currents from the utility supply. If the supply impedance is relatively high significant distortion of the mains voltage waveform is inevitable unless filters are fitted on the AC input side, but with normal industrial supplies there is no problem for small inverters of a few kW rating.

2. Steady-State Operation—Importance of Achieving Full Flux

Three simple relationships need to be borne in mind to simplify understanding of how the inverter-fed induction motor behaves. Firstly, we established that for a given induction motor, the torque developed depends on the strength of the rotating flux density wave, and on the slip speed of the rotor, i. e. on the relative velocity of the rotor with respect to the flux wave. Secondly, the strength or amplitude of the flux wave depends directly on the supply voltage to the stator windings, and inversely on the supply frequency. And thirdly, the absolute speed of the flux wave depends directly on the supply frequency.

Recalling that the motor can only operate efficiently when the slip is small, we see that the basic method of speed control rests on the control of the speed of rotation of the flux wave (i. e. the synchronous speed), by control of the supply frequency. If the motor is a 4-pole one, for example, the synchronous speed will be 1,500r/min when supplied at 50Hz, 1,200r/min at 40Hz, 750r/min at 25Hz and so on. The no-load speed will therefore be almost exactly proportional to the supply frequency, because the torque at no load is small and the corresponding slip is also very small.

Turning now to what happens on load, we know that when a load is applied the rotor slows down, the slip increases, more current is induced in the rotor, and more torque is produced.[10] When the speed has reduced to the point where the motor torque equals the load torque, the speed becomes steady. We normally want the drop in speed with load to be as small as possible, not only to minimise the drop in speed with load, but also to maximise efficiency: in short, we want to minimise the slip for a given load.

It is known that the slip for a given torque depends on the amplitude of the rotating flux wave: the higher the flux, the smaller the slip needed for a given torque. It follows that having set the desired speed of rotation of the flux wave by controlling the output frequency of the inverter we must also ensure that the magnitude of the flux is adjusted so that it is at its full (rated) value, regardless of the speed of rotation. This is achieved by making the output voltage from the inverter vary in the appropriate way in relation to the frequency.

We recall that the amplitude of the flux wave is proportional to the supply voltage and inversely proportional to the frequency, so if we arrange that the voltage supplied by the inverter vary in direct proportion to the frequency, the flux wave will have a constant amplitude. This philosophy is at the heart of most inverter-fed drive systems: there are variations, as we will see, but in the majority of cases the internal control of the inverter will be designed so that the output voltage to frequency ratio (V/f) is automatically kept constant, at least up to the 'base' (50Hz or 60Hz) frequency.

Many inverters are designed for direct connection to the mains supply, without a transformer, and as a result the maximum inverter output voltage is limited to a value similar to that of the mains. With a 415V supply, for example, the maximum inverter output voltage will be perhaps 450V. Since the inverter will normally be used to supply a standard induction motor designed for say 415V, 50Hz operation, it is obvious that when the inverter is set to deliver 50Hz, the voltage should be 415V, which is within the inverter's voltage range. But when the frequency was raised to say 100Hz, the voltage should-ideally-be increased to 830V in order to obtain full flux. The inverter cannot supply voltages above 450V, and it follows that in this case full flux can only be maintained up to speeds a little above base speed.

Established practice is for the inverter to be capable of maintaining the V/f ratio constant up to the base speed, but to accept that at all higher frequencies the voltage will be constant at its maximum value.[11] This means that the flux is maintained constant at speeds up to base speed, but beyond that the flux reduces inversely with frequency. Needless to say the performance above base speed is adversely affected, as we will see.

Users are sometimes alarmed to discover that both voltage and frequency change when a new speed is demanded. Particular concern is expressed when the voltage is seen to reduce when a lower speed is called for. Surely, it is argued, it can't be right to operate say a 400V induction motor at anything less than 400V. The fallacy in this view should now be apparent: the figure of 400V is simply the correct voltage for the motor when run directly from the mains, at say 50Hz. If this full voltage was applied when the frequency was reduced to say 25Hz, the implication would be that the flux would have to rise to twice its rated value.

This would greatly overload the magnetic circuit of the machine, giving rise to excessive saturation of the iron, an enormous magnetising current and wholly unacceptable iron and copper losses. To prevent this from happening, and keep the flux at its rated value, it is essential to reduce the voltage in proportion to frequency. In the case above, for example, the correct voltage at 25Hz would be 200V.

7.2.2 Torque-Speed V/F Operation Constant

When the voltage at each frequency is adjusted so that the ratio V/f is kept constant up to base speed, and full voltage is applied thereafter, a family of torque-speed curves as shown in Fig. 7-7 is obtained. These curves are typical for a standard induction motor of several kW output.

As expected, the no-load speeds are directly proportional to the frequency, and if the frequency is held constant, e.g. at 25Hz in Fig. 7-7, the speed drops only modestly from no-load (point a) to full-load (point b). These are therefore good open-loop characteristics, because the speed is held fairly well from no-load to full-load. If the application calls for the speed to be held precisely, this can clearly be achieved (with the aid of closed-loop speed control) by raising the frequency so that the full-load operating point moves to point c.

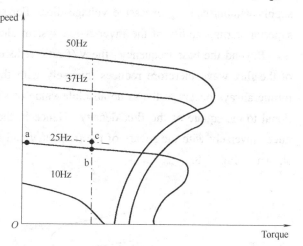

Fig. 7-7 Torque-Speed Curves for Inverter-Fed Induction Motor with Constant Voltage-Frequency Ratio

We also note that the pull-out torque and the torque stiffness (i.e. the slope of the torque-speed curve in the normal operating region) is more or less the same at all points below base speed, except at low frequencies where the effect of stator resistance in reducing the flux becomes very pronounced. It is clear from Fig. 7-7 that the starting torque at the minimum frequency is much less than the pull-out torque at higher frequencies, and this could be a problem for loads which require a high starting torque.

The low-frequency performance can be improved by increasing the V/f ratio at low frequencies in order to restore full flux, a technique which is referred to as 'low-speed voltage boosting'. Most drives incorporate provision for some form of voltage boost, either by way of a single adjustment to allow the user to set the desired starting torque, or by means of more complex provision for varying the V/f ratio over a range of frequencies.

This region of the characteristics is known as the 'constant torque' region, which means that for frequencies up to base speed, the maximum possible torque which the motor can deliver is independent of the set speed. Continuous operation at peak torque will not be allowable because the motor will overheat, so an upper limit will be imposed by the controller, as discussed shortly. [12] With this imposed limit, operation below base speed corresponds to the armature-voltage control region of a DC drive.

We should note that the availability of high torque at low speeds (especially at zero speed) means that we can avoid all the 'starting' problems associated with fixed-frequency operation. By starting off with a low frequency which is then gradually raised the slip speed of the rotor is always small, i. e. the rotor operates in the optimum condition for torque production all the time, thereby avoiding all the disadvantages of high-slip (low torque and high current) that are associated with mains-frequency starting. This means that not only can the inverter-fed motor provide rated torque at low speeds, but perhaps more importantly—it does so without drawing any more current from the mains than under full-load conditions, which means that we can safely operate from a weak supply without causing excessive voltage dips. For some essentially fixed-speed applications, the superior starting ability of the inverter-fed system alone may justify its cost.

Beyond the base frequency, the V/f ratio reduces because V remains constant. The amplitude of the flux wave therefore reduces inversely with the frequency. Now we saw that the pull-out torque always occurs at the same absolute value of slip speed, and that the peak torque is proportional to the square of the flux density. Hence in the constant voltage region the peak torque reduces inversely with the square of the frequency and the torque-speed curve becomes less steep, as shown in Fig. 7-8.

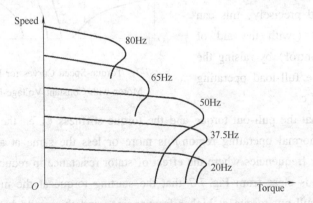

Fig. 7-8 Typical Torque-Speed Curves for Inverter-Fed Induction Motor with Low-Speed Voltage Boost, Constant Voltage-Frequency Ratio from Low Speed up to Base Speed, and Constant Voltage above Base Speed

7.2.3 Vector (Field-Oriented) Control

We have seen previously that in both the induction motor and the DC motor, torque is produced by the interaction of currents on the rotor with the radial flux density produced by the stator. Thus to change the torque, we must either change the magnitude of the flux, or the rotor current, or both; and if we want a sudden (step) increase in torque, we must make the change (or changes) instantaneously.

Since every magnetic field has stored energy associated with it, it should be clear that it is not possible to change a magnetic field instantaneously, as this would require the energy to change in zero time, which calls for a pulse of infinite power. In the case of the main field of a motor, we could not hope to make changes fast enough even to approximate the step change in torque we are seeking, so the only alternative is to make the rotor current change as quickly as possible.

In the DC motor it is relatively easy to make very rapid changes in the armature (rotor) current because we have direct access to the armature current via the brushes. The armature circuit inductance is relatively low, so as long as we have plenty of voltage available, we can apply a large voltage (for a very short time) whenever we want to make a sudden change in the armature current and torque. This is done automatically by the inner (current-control) loop in the DC drive.

In the induction motor, matters are less straightforward because we have no direct access to the rotor currents, which have to be induced from the stator side. Nevertheless, because the stator and rotor windings are tightly coupled via the air-gap field, it is possible to make more or less instantaneous changes to the induced currents in the rotor, by making instantaneous changes to the stator currents. [13] Any sudden change in the stator MMF pattern (resulting from a change in the stator currents) is immediately countered by an opposing rotor MMF set up by the additional rotor currents which suddenly spring up. All tightly coupled circuits behave in this way, the classic example being the transformer, in which any sudden change in say the secondary current is immediately accompanied by a corresponding change in the primary current. Organising these sudden step changes in the rotor currents represents both the essence and the challenge of the vector-control method.

We have already said that we have to make sudden step changes in the stator currents, and this is achieved by providing each phase with a fast-acting closed-loop current controller. Fortunately, under transient conditions the effective inductance looking in at the stator is quite small (it is equal to the leakage inductance), so it is possible to obtain very rapid changes in the stator currents by applying high, short-duration impulsive voltages to the stator windings. In this respect each stator current controller closely resembles the armature current controller used in the DC drive. [14]

When a step change in torque is required, the magnitude, frequency and phase of the three stator currents are changed (almost) instantaneously in such a way that the frequency, magnitude and phase of the rotor current wave jump suddenly from one steady state to another. This change is done without altering the amplitude or position of the resultant rotor flux linkage relative to the rotor, i. e. without altering the stored energy significantly. The flux density term (B) therefore remains the same while the terms I_r and φ_r change instantaneously to their new steady-state values, corresponding to the new steady-state slip and torque.

We can picture what happens by asking what we would see if we were able to observe the stator MMF wave at the instant that a step increase in torque was demanded. For the sake of simplicity, we will assume that the rotor speed remains constant, in which case we would find that:

1) The stator MMF wave suddenly increases its amplitude.

2) It suddenly accelerates to a new synchronous speed.

3) It jumps forward to retain its correct relative phase with respect to the rotor flux and current waves.

Thereafter the stator MMF retains its new amplitude, and rotates at its new speed. The rotor experiences a sudden increase in its current and torque, the new current being maintained by the new (higher) stator currents and slip frequency.

We should note that both before and after the sudden changes, the motor operates in the normal fashion, as discussed earlier. The 'vector control' is merely the means by which we are able to make a sudden stepwise transition from one steady state operating condition to another, and it has no effect whatsoever once we have reached the steady state.

The unique feature of the vector drive which differentiates it from the ordinary or scalar drive (in which only the magnitude and frequency of the stator MMF wave changes when more torque is required) is that by making the right sudden change to the instantaneous position of their stator MMF wave, the transition from one steady state to the other is achieved instantaneously, without the variables hunting around before settling to their new values. In particular, the vector approach allows us to overcome the long electrical time-constant of the rotor cage, which is responsible for the inherently sluggish transient response of the induction motor. It should also be pointed out that, in practice, the speed of the rotor will not remain constant when the torque changes so that, in order to keep track of the exact position of the rotor flux wave, it will be necessary to have a rotor position feedback signal.

Because the induction motor is a multi-variable non-linear system, an elaborate mathematical model of the motor is required, and implementation of the complex control algorithms calls for a large number of fast computations to be continually carried out so that the right instantaneous voltages are applied to each stator winding. This has only recently been made possible by using sophisticated and powerful signal processing in the drive control.

No industry standard approach to vector control has yet emerged, but systems fall into two broad categories, depending on whether or not they employ feedback from a shaft-mounted encoder to track the instantaneous position of the rotor. Those that do are known as 'direct' methods, whereas those which rely entirely on a mathematical model of the motor are known as 'indirect' methods. Both systems use current feedback as an integral part of each stator current controllers, so at least two stator current sensors are required.[15] Direct systems are inherently more robust and less sensitive to changes in machine parameters, but call for a non-standard (i.e. more expensive) motor and encoder.

The dynamic performance of direct vector drives is now so good that they are found in demanding roles that were previously the exclusive preserve of the DC drive, such as reversing drives and positioning applications. The achievement of such outstandingly impressive performance from a motor whose inherent transient behaviour is poor represents a major milestone in the already impressive history of the induction motor.

New Words and Expressions

1. mercury arc 汞弧
2. regeneration *n.* 再生
3. yardstick *n.* 尺码，准绳
4. firing angle 触发角
5. laminated *adj.* 分层的，叠片的
6. blower *n.* 鼓风机，融固器
7. ventilation *n.* 通风
8. potentiometer *n.* 电位器
9. off-the-shelf 成品，通用件
10. inertia *n.* 惯性，惯量
11. chunk *n.* 组块，大量
12. tachogenerator *n.* 测速发电机
13. ambient *n.* 周围，环境
14. reference signal 参考信号
15. rated current 额定电流
16. magnitude *n.* 幅度，大小
17. judicious *adj.* 明智的
18. mark-space 脉冲间隔
19. predominant *adj.* 有优势的
20. phase-locked loop 锁相环
21. inverter-fed induction motor 逆变器供电的感应电动机
22. tachometer *n.* 转速表
23. main frequency 主频率
24. instantaneous *adj.* 瞬间的，即刻的
25. from instant to instant 时时刻刻
26. bulky *adj.* 大的，体积大的
27. electrolytic capacitor 电解电容
28. current source inverters (CSI) 电流源型变换器
29. harmonic *n.* 谐波
30. acoustic noise 噪声
31. V/f ratio 压频比
32. fallacy *n.* 谬误，谬论
33. pull-out 活页
34. vector control 矢量控制
35. compare unfavourably with 相对劣于
36. direct-on-line start 直接启动
37. quasi-steady-state 准稳态
38. run-up 前奏，预备阶段
39. coupled circuits 耦合电路
40. MMF wave 磁动势

Notes

[1] The generator was driven at fixed speed by an induction motor, and the field of the generator was varied in order to vary the generated voltage.

发电机由定速感应电动机驱动，为了改变发电机的电压，其励磁是变化的。

[2] By altering the firing angle of the thyristors the mean value of the rectified voltage can be varied, thereby allowing the motor speed to be controlled.

通过改变触发角来改变整流电压的平均值，从而控制电动机的转速。

[3] The voltage waveform can be considered to consist of a mean DC level (U_d), and a superimposed pulsating or ripple component.

电压波形可以认为是由一个平均直流电压（U_d）和叠加在其上的脉动或纹波成分组成的。

[4] The ripple voltage causes a ripple current to flow in the armature, but because of the armature inductance, the amplitude of the ripple current is small.

纹波电压引起纹波电流流过电枢，但由于电枢电感的存在，纹波电流的幅值很小。

[5] The speed error is amplified by the speed controller (which is more accurately described

as a speed-error amplifier) and the output serves as the reference or input signal to the inner control system.

转速调节器将转速误差放大（更确切地说是转速误差放大器），其输出作为内环控制的给定或输入信号。

[6] A single-switch chopper using a transistor, MOSFET or IGBT can only supply positive voltage and current to a DC motor, and is therefore restricted to quadrant 1 motoring operation.

一个开关的斩波器使用一个晶体管、MOSFET 或 IGBT，只能给直流电动机提供正电压和正电流，因而（电动机）只能运行在第一象限。

[7] The shape of the armature voltage waveform reminds us that when the transistor is switched on, the battery voltage V is applied directly to the armature, and for the remainder of the cycle the transistor is turned off and the current freewheels through the diode.

从电枢电压的波形可以看出，当晶体管开关器件导通时，电源电压 V 直接加到电枢上，而在开关器件关断的周期内，电流将通过二极管续流。

[8] The control philosophy and arrangements for a chopper-fed motor are the same as for the converter-fed motor, with the obvious exception that the mark-space ratio of the chopper is used to vary the output voltage, rather than the firing angle.

直流电动机斩波调压的控制原理和方案与晶闸管整流调压相似，只有一个明显的差别，那就是，斩波器是通过改变占空比而不是改变触发角来改变输出电压的。

[9] They can create unpleasant acoustic noise, and always give rise to additional iron and copper losses.

它们会产生讨厌的噪声，并且会产生额外的铁耗和铜耗。

[10] Turning now to what happens on load, we know that when a load is applied the rotor slows down, the slip increases, more current is induced in the rotor, and more torque is produced.

现在来看看负载时的情况，我们知道，负载时转子会慢下来，转差将变大，转子感应出更大的电流，于是产生更大的转矩。

[11] Established practice is for the inverter to be capable of maintaining the V/f ratio constant up to the base speed, but to accept that at all higher frequencies the voltage will be constant at its maximum value.

按照惯例，变频器应能保证在基频以下 V/f 比为常数，而在基频以上电压将保持为最大值不变。

[12] Continuous operation at peak torque will not be allowable because the motor will overheat, so an upper limit will be imposed by the controller, as discussed shortly.

电动机不可以连续运行在最大转矩下，这样将导致电动机过热，因此将由调节器限定其上限值，如前简述。

[13] Nevertheless, because the stator and rotor windings are tightly coupled via the air-gap field, it is possible to make more or less instantaneous changes to the induced currents in the rotor, by making instantaneous changes to the stator currents.

然而，由于定子和转子是通过气隙紧密耦合的，通过定子电流的瞬时变化，使转子感应的电流产生或多或少的瞬时变化还是可能的。

[14] In this respect each stator current controller closely resembles the armature current controller used in the DC drive.

在这个方面，每个定子电流调节器很像直流电动机中的电枢电流调节器。

[15] Both systems use current feedback as an integral part of each stator current controllers, so at least two stator current sensors are required.

（由于）两个系统中定子电流调节器都采用了电流反馈，因此至少需要两个定子电流传感器。

Chapter 8 Programmable Logic Controller Technology

8.1 Introduction

Programmable Logic Controller (PLC) is actually an industrial micro-controller system (in more recent times we meet processors instead of micro-controllers) where you have hardware and software specifically adapted to industrial environment. It is a device that was invented to replace the necessary sequential relay circuits for machine control. [1] The PLC works by looking at its inputs and depending upon their state, turning on/off its outputs. The user enters a program, usually via software or programmer that gives the desired results.

8.1.1 Central Processing Unit

Central Processing Unit (CPU) is the brain of a PLC controller. CPU itself is usually one of the micro-controllers. In the past these were 8-bit micro-controllers such as 8051, and now these are 16-bit and 32-bit micro-controllers. Unspoken rule is that you will find mostly Hitachi and Fujicu micro-controllers in PLC controllers by Japanese makers, Siemens in European controllers, and Motorola micro-controllers in American ones. [2] CPU also takes care of communication, interconnection among other parts of PLC controller, program execution, memory operation, overseeing input and setting up of an output. PLC controllers have complex routines for memory checkup in order to ensure that PLC memory is not damaged (memory checkup is done for safety reasons). Generally speaking, CPU unit makes a great number of check-ups of the PLC controller itself so eventual errors would be discovered early. You can simply look at any PLC controller and see that there are several indicators in the form of light diodes for error signalization.

8.1.2 Memory

System memory (today mostly implemented in FLASH technology) is used by a PLC for an process control system. Aside from this operating system it also contains a user program translated from a ladder diagram to a binary form. FLASH memory contents can be changed only in case where user program is being changed. PLC controllers were used earlier instead of FLASH memory and have had EPROM memory instead of FLASH memory which had to be erased with UV lamp and programmed on programmers. [3] With the use of FLASH technology this process is greatly shortened. Reprogramming a program memory is done through a serial cable in a program for application development. Use memory is divided into blocks having special functions. Some parts of a memory are used for storing input and output status. The real status of an input is stored either as "1" or as "0" in a specific memory bit. Each input or output has one corresponding bit in memory. Other parts of memory are used to store variable contents for variables used in user program. For example, timer value, or counter value would be stored in this part of the memory.

8.1.3 Programming a PLC Controller

PLC controller can be reprogrammed through a computer (usual way), but also through manual programmers (consoles). This practically means that each PLC controller can be programmed through a computer if you have the software needed for programming. Today's transmission computers are ideal for reprogramming a PLC controller in factory itself. This is of great importance to industry. Once the system is corrected, it is also important to read the right program into a PLC again. It is also good to check from time to time whether program in a PLC has not changed. This helps to avoid hazardous situations in factory rooms (some automakers have established communication networks which regularly check programs in PLC controllers to ensure execution only of good programs).[4] Almost every program for programming a PLC controller possesses various useful options such as: forced switching on and off of the system inputs/outputs (I/O lines), program follow up in real time as well as documenting a diagram. This documenting is necessary to understand and define failures and malfunctions. Programmer can add remarks, names of input or output devices, and comments that can be useful when finding errors, or with system maintenance. Adding comments and remarks enables any technician (and not just a person who developed the system) to understand a ladder diagram right away. Comments and remarks can even quote precisely part number if replacements would be needed. This would speed up a repair of any problems that come up due to bad parts. The old way was such that a person who developed a system had protection on the program. So nobody aside from this person could understand how it was done. Correctly documented ladder diagram allows any technician to understand thoroughly how system functions.

8.1.4 Power Supply

Electrical supply is used in bringing electrical energy to central processing unit. Most PLC controllers work either at 24V DC or 220V AC. On some PLC controllers you will find electrical supply as a separate module. Those are usually bigger PLC controllers, while small and medium series already contain the supply module. User has to determine how much current to take from I/O module to ensure that electrical supply provides appropriate amount of current. Different types of modules use different amounts of electrical current. This electrical supply is usually not used to start external inputs or outputs. User has to provide separate supplies in starting PLC controller inputs or outputs because then you can ensure so called "pure" supply for the PLC controller.[5] With pure supply we mean supply where industrial environment can not affect it damaged. Some of the smaller PLC controllers supply their inputs with voltage from a small supply source already incorporated into a PLC.

8.1.5 PLC Controller Inputs

Intelligence of an automated system depends largely on the ability of a PLC controller to read signals from different types of sensors and input devices. Keys, keyboards and functional switches are a basis for man versus machine relationship. On the other hand, in order to detect a working piece, view a mechanism in motion, check pressure or fluid level you need specific automatic devices such as proximity sensors, marginal switches, photoelectric sensors, level sensors, etc. [6] Thus, input signals can be logical (on/off) or analogue. Smaller PLC controllers usually have only digital input lines while larger also accept analogue inputs through special units attached to PLC controller. One of the most frequent analogue signals is a current signal of 4 mA to 20 mA and mega-voltage signal generated by various sensors. Sensors are usually used as inputs for PLC. You can obtain sensors for different purposes. They can sense presence of some parts, measure temperature, pressure, or some other physical dimension, etc. (ex. inductive sensors can register metal objects). Other devices also can serve as inputs to PLC controller. Intelligent devices such as robots, video system, etc. often are capable of sending signals to PLC controller input modules (robot, for instance, can send a signal to PLC controller inputs as information when it has finished moving an object from one place to the other).

8.1.6 Input Adjustment Interface

Adjustment interface also called an interface is placed between input lines and a CPU unit. The purpose of adjustment interface is to protect a CPU from disproportionate signals from an outside world. Input adjustment module turns a level of real logic to a level that suits CPU unit (ex. input from a sensor which works on 24V DC must be converted to a signal of 5V DC in order for a CPU to be able to process it). This is typically done through opto-isolation, and this function you can view in Fig. 8-1. Opto-isolation means that there is no electrical connection between external world and CPU unit. They are "optically" separated, or in other words, signal is transmitted through light. [7]

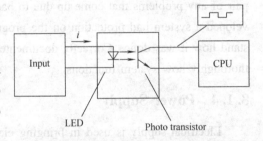

Fig. 8-1 Input Adjustable Interface

The way this works is simple. External device brings a signal which turns LED on, whose light in turn incites photos transistor which in turn starts conducting, and a CPU sees this as logic zero (supply between collector and transmitter falls under 1V). When input signal stops LED diode turns off, transistor stops conducting, collector voltage increases, and CPU receives logic 1 as information.

8.1.7 PLC Controller Output

Automated system is incomplete if it is not connected with some output devices. Some of the most frequently used devices are motors, solenoids, relays, indicators, sound signalization and similar. By starting a motor, or a relay, PLC can manage or control a simple system such as system for sorting products all the way up to complex systems such as service system for positioning head of CNC machine. Output can be of analogue or digital type. Digital output signal works as a switch and it connects and disconnects line. Analogue output is used to generate the analogue signal (ex. motor whose speed is controlled by a voltage that corresponds to a desired speed).

8.1.8 Output Adjustment Interface

Output interface is similar to input interface, as shown in Fig. 8-2. CPU brings a signal to LED diode and turns it on. Light incites a photo transistor which begins to conduct electricity, and thus the voltage between collector and emitter falls to 0.7V, and a device attached to this output sees this as a logic zero.

Inversely it means that a signal at the output exists and is interpreted as logic one. Photo transistor is not directly connected to a PLC controller output. Between photo transistor and an output usually there is a relay or a stronger transistor capable of interrupting stronger signals.

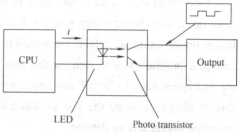

Fig. 8-2 Output Adjustable Interface

8.1.9 Extension Lines

Every PLC controller has a limited number of input/output lines. If needed this number can be increased through certain additional modules by system extension through extension lines. Each module can contain extension both of input and output lines. Also, extension modules can have inputs and outputs of a different nature from those on the PLC controller (ex. in case relay outputs are on a controller, transistor outputs can be on an extension module).

8.2 PLC Operation Process

A PLC works by continually scanning a program. We can think of this scan cycle as consisting of three important steps, as shown in Fig. 8-3. There are typically more than three but we can focus on the important parts and not worry about the others. Typically the others are checking the system and updating the current internal counter and timer values.

Step 1—Check Input Status

First the PLC takes a look at each input to determine if it is on or off. In other words, is the sensor connected to the first input on? How about the second input? How about the third? It records this data into its memory to be

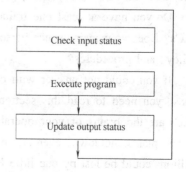

Fig. 8-3 The Work Process of a PLC

used during the next step.

Step 2—Execute Program

Next the PLC executes your program on instruction at a time. Maybe your program said that if the first input was on then it should turn on the first output. Since it already knows which inputs are on/off from the previous step, it will be able to decide whether the first output should be turned on based on the state of the first input. [8] It will store the execution results for use later during the next step.

Step 3—Update Output Status

Finally the PLC updates the status of the outputs. It updates the outputs based on which inputs were on during the first step and the results of executing your program during the second step. Based on the example in Step 2 it would now turn on the first output because the first input was on and your program said to turn on the first output when this condition is true. [9] After the Step 3 the PLC goes back to Step 1 and repeats the steps continuously. One scan time is defined as the time it takes to execute the 3 steps listed above. Thus a practical system is controlled to perform specified operations as desired.

Since technology for motion control of electric drives became available, the use of programmable logic controllers (PLCs) with power electronics in electric machines applications has been introduced in the manufacturing automation. [10] This use offers advantages such as lower voltage drop when turned on and the ability to control motors and other equipment with a virtually unity power factor. Many factories use PLCs in automation processes to diminish production cost and to increase quality and reliability. Other applications include machine tools with improved precision computerized numerical control (CNC) due to the use of PLCs. To obtain accurate industrial electric drive systems, it is necessary to use PLCs interfaced with power converters, personal computers, and other electric equipment. Nevertheless, this makes the equipment more sophisticated, complex, and expensive.

8.3 PLC Maintenance Management

What is a PLC? How many PLCs is your bottom line depending on? Do you have an up to date list of all PLC model types, part availability, program copies, and details for your company?

Do you have at least one trained person per shift, to maintain and troubleshoot your plant PLCs? Does your maintenance personnel work with PLCs following written company or corporate policy, and procedures?

If you could not answer with confidence or you answered 'No' to any of the above questions, you need to read this section on maintenance management of PLCs. Why? Because the PLCs are the brains of your operation. When the PLC is not functioning properly, lines shut down, plants shutdown, even city bridges and water stations could cease to operate. Thousands to millions could be lost by one little PLC in an electrical panel that you never even knew existed. But most importantly, damage to machine and personnel could result from improper maintenance management of your company's PLCs. [11]

8.3.1 What Is A PLC

First, What a PLC is? As this section is not just for the maintenance technician, but for maintenance managers, plant managers and corporate managers. A PLC is the type of computer that controls most machines today. The PLC is used to control and to troubleshoot the machine. The PLC is the brain of the machine. Without it, the machine is dead. The maintenance technicians we train are the brain surgeons.

Important Note: Just as a doctor asks the patient questions to figure out what is wrong, a maintenance technician asks the PLC questions to troubleshoot the machine. [12] The maintenance technician uses a laptop computer to see what conditions have to be met in order for the PLC to cause an action to occur (like turn a motor on). In a reliable maintenance management environment, the maintenance technician will be using the PLC as a troubleshooting tool to reduce downtime.

A little more detailed definition of a PLC: A programmable controller is a small industrial strength computer used to control real world actions, based on its program and real world sensors. The PLC replaces thousands of relays that were in older electrical panels, and allows the maintenance technician to change the way a machine works without having to do any wiring. The program is typically in ladder logic, which is similar to the wiring schematics maintenance electricians are already accustomed to working with. Inputs to a PLC can be switches, sensors, bar codes, machine operator data, etc. Outputs from the PLC can be motors, air solenoids, indicator lights, etc.

8.3.2 How Many PLCs Is Your Bottom Line Depending on

It is common to only learn about a PLC once the machine is down and the clock is ticking at a thousand dollars an hour, or more. [13] Unfortunately, it is also common that after the fire is out, it is on to the next fire, without fully learning what can be done to avoid these costly downtime in the future, and in other similar machines in a company or corporation.

Some older electrical panels may only have relays in them, but most machines are controlled by a PLC. A bottleneck machine in your facility may have a PLC. Most plant air compressors have a PLC. How much would it cost if the bottleneck or plant air shut down a line, a section of your facility, or even the entire plant?

8.3.3 Do You Have an Up to Date List of All PLC for Your Company

The first step to take is to perform a PLC audit. Open every electrical panel, and write down the PLC brand, model, and other pertinent information. Then go the next two steps. Analyze the audit information and risk, then act on that analysis.

Once you have collected the basic information in your plant wide and/or corporate audit, you need to analyze the information to develop an action plan based on risk analysis. In the risk analysis, bottlenecks and other factors will help you assess priorities. Starting with the highest priority PLC, you will need to ask more important questions.

1) Do we have the most common spares for the PLC?

2) Is the OEM (Original Equipment Manufacturer) available? Or even in business any more?

3) Do we have a back up copy of the PLC program?

4) Does our program copy have descriptions so we can work with it reliably and efficiently?

5) Do we have the software needed to view the PLC program? Are our maintenance personnel trained on that PLC brand?

These are some of the questions our managers must ask, to avoid unnecessary risk and to insure reliability.

8.3.4 Do You Have Trained Person Per Shift to Maintain and Troubleshoot Your Plant PLCs

Is your maintenance staff trained on the PLC? (Silly to squander over a couple thousand in maintenance training when the lack of PLC knowledge could cost you 10 thousand an hour... or worse.) There are a couple good reasons why you should have at least one trained person per shift, to work reliably with PLCs. You do not want to see greater downtime on off shifts because the knowledge base is on day shift only. Also with all the baby boomers (our core knowledge base in the industry) about to retire, it is not smart management to place all your eggs in one basket. [14] Then the question should be asked, what should we look for in training? You should seek training with two primary objectives.

1) Firstly, the training you decide on, should stress working with PLCs in a Safe and Reliable way (not just textbook knowledge or self-learned knowledge).

2) Secondly, the training should be actually centered around the PLC products you are using or plan to use in your facility.

Some other good ideas to get more out of your PLC training investment would be to get hands on training using the actual PLC programs and software the maintenance technician will be working with in the facility. [15] Insure your personnel have the software, equipment and encouragement to continue with self-education. PLC Training CBT (Computer Based Training) CDs are a great way for employees to follow up 6 months after the initial training. Some other ideas you could do is to provide them with simulation software and/or a spare PLC off the shelf to practice with.

8.4 The Application Future of PLC

During the three decades following their introduction, PLCs have evolved to incorporate analog I/O, communication over networks, and new programming standards. [16] However, engineers create 80 percent of industrial applications with digital I/O, a few analog I/O points, and simple programming techniques. Experts ARC, Venture Development Corporation and the online PLC training source PLCS. net estimate that: 77% of PLCs are used in small applications (less than 128 I/O); 72% of PLC I/O is digital; 80% of PLC application challenges are solved with a set of 20 ladder-logic instructions.

Because 80 percent of industrial applications are solved with traditional tools, there is strong demand for simple low-cost PLCs. This has spurred the growth of low-cost micro PLCs with digital I/O that use ladder logic. It has also created a discontinuity in controller technology, where 80 percent of applications require simple, low cost controllers and 20 percent relentlessly push the capabilities of traditional control systems. The applications that fall within the 20 percent are built by engineers who require higher loop rates, advanced control algorithms, more analogy capabilities, and better integration with the enterprise network. [17]

In the 1980s and1990s, these "20 percenters" evaluated PCs for industrial control. The PC provided the software capabilities to perform advanced tasks, offered a graphical rich programming and user environment, and utilized COTS components allowing control engineers to take advantage of technologies developed for other applications. These technologies include floating point processors; high speed I/O buses, such as PCI and Ethernet; non-volatile data storage; and graphical development software tools. The PC also provided unparalleled flexibility, highly productive software, and advanced low-cost hardware.

However PCs were still not ideal for control applications. Although many engineers used the PC when incorporating functionality, such as analogy control and simulation, database connectivity, web based functionality, and communication with third party devices, the PLC still ruled the control realm. The main problem with PC-based control was that standard PCs were not designed for rugged environments. The PC presented three main challenges. Often, the PCs general-purpose operating system was not stable enough for control. PC-controlled installations were forced to handle system crashes and unplanned rebooting. With rotating magnetic hard drives and non-hardened components, such as power supplies, PCs were more prone to failure. Plant operators need the ability to override a system for maintenance or troubleshooting. Using ladder logic, they can manually force a coil to a desired state, and quickly patch the affected code to quickly override a system. However, PC systems require operators learn new, more advanced tools.

Although some engineers use special industrial computers with rugged hard ware and special operating systems, most engineers avoided PCs for control because of problems with PC reliability. In addition, the devices used within a PC for different automation tasks, such as I/O, communications or motion, may have different development environments. So the "20 percenters" either lived without functionality not easily accomplished with a PLC or cobbled together a system that included a PLC for the control portion of the code and a PC for the more advanced functionality. [18]

This is the reason many factory floors today have PLCs used in conjunction with PCs for data logging, connecting to bar code scanners, inserting information into databases, and publishing data to the Web. The big problem with this type of setup is that these systems are often difficult to construct, troubleshoot and maintain. The system engineer often is left with the unenviable task of incorporating hardware and software from multiple vendors, which poses a challenge because the equipment is not designed to work together.

With no clear PC or PLC solution, engineers with complex applications worked closely with control vendors to develop new products. [19] They requested the ability to combine the advanced software capabilities of the PC with the reliability of the PLC. These lead users helped guide product development for PLC and PC-based control companies.

The software capabilities required not only advanced software, but also an increase in the hardware capabilities of the controllers. With the decline in world-wide demand for PC components, many semiconductor vendors began to redesign their products for industrial applications. Control vendors today are incorporating industrial versions of floating point processors, DRAM, solid-state storage devices such as Compact Flash, and fast Ethernet chipsets into industrial control products. This enables vendors to develop more powerful software with the flexibility and usability of PC-based control systems that can run on real-time operating systems for reliability. [20]

The resulting new controllers, designed to address the "20 percent" applications, combine the best PLC features with the best PC features. Industry analysts at ARC named these devices programmable automation controllers, or PACs. In their "Programmable Logic Controllers Worldwide Outlook" study, ARC identified five main PAC characteristics. These criteria characterize the functionality of the controller by defining the software capabilities.

New Words and Expressions

1. relay   *n.* 继电器
2. circuit   *n.* 电路
3. ladder diagram   梯形图
4. CPU   中央处理单元
5. bit   *n.* 位
6. memory   *n.* 内存
7. diode   *n.* 二极管
8. cable   *n.* 电报，电缆
9. timer   *n.* 定时器
10. counter   *n.* 计数器
11. console   *n.* 控制台
12. voltage   *n.* 电压
13. sensor   *n.* 传感器
14. analogue   *n.* 模拟，类似物
15. digital   *adj. & n.* 数字（的）
16. pressure   *n.* 压力
17. robot   *n.* 机器人，机械手
18. level   *n.* 水平，电平
19. opto-isolation   光隔离
20. transistor *n.* 晶体管

21. motor   *n.* 电动机，马达
22. solenoid   *n.* 电磁阀
23. scan cycle 扫描周期
24. maintenance *n.* 维修，保持，维护
25. procedure   *n.* 程序，步骤，规程
26. troubleshoot *v.* 故障排除
27. laptop   *n.* 便携式计算机
28. sensor   *n.* 传感器
29. bar code   *n.* 条形码
30. bottleneck *n.* 瓶颈 *v.* 使……为难
31. network node   网络节点
32. model number   型号
33. spare   *adj.* 备用的 *n.* 备件
34. back up   备份
35. reliability *n.* 可靠性
36. downtime   *n.* 故障停机时间
37. criteria   *n.* 标准
38. simulation software   仿真软件
39. floppy disk   软盘
40. filter   *n.* 过滤器，滤波器

Notes

[1] It is a device that was invented to replace the necessary sequential relay circuits for machine control.

它是为替代实现机器顺序控制所必需的继电器电路而发明的装置。

[2] Unspoken rule is that you will find mostly Hitachi and Fujicu micro-controllers in PLC controllers by Japanese makers, Siemens in European controllers, and Motorola micro-controllers in American ones.

厂家应用的潜规则是，日本制造商使用的是日立和富士通微控制器，欧洲制造商使用西门子控制器，美国制造商使用的是摩托罗拉微控制器。

[3] PLC controllers were used earlier instead of FLASH memory and have had EPROM memory instead of FLASH memory which had to be erased with UV lamp and programmed on programmers.

早期 PLC 控制器用来代替闪存，用 EPROM 内存代替闪存，它必须用紫外线灯和编辑器中的程序进行擦除。

[4] This helps to avoid hazardous situations in factory rooms (some automakers have established communication networks which regularly check programs in PLC controllers to ensure execution only of good programs).

这有助于工厂车间内避免发生危险情况（一些汽车制造商已经建立了通信网络，可以通过 PLC 控制器定期检查程序以确保运行程序的正确性）。

[5] User has to provide separate supplies in starting PLC controller inputs or outputs because then you can ensure so called "pure" supply for the PLC controller.

用户在启动 PLC 控制器输入或输出时必须提供独立的电源，因为这样你可以确保给 PLC 控制器提供所谓的"纯净"电源。

[6] On the other hand, in order to detect a working piece, view a mechanism in motion, check pressure or fluid level you need specific automatic devices such as proximity sensors, marginal switches, photoelectric sensors, level sensors, etc.

另一方面，为了监测一个工作过程，观测一个机械运动，检查压力或液位需要特定的自动化设备来完成，如近程传感器、临界开关、光电传感器、液位传感器等。

[7] They are "optically" separated, or in other words, signal is transmitted through light.

它们是"光"分离，或者换句话说，信号通过光进行传递。

[8] Since it already knows which inputs are on/off from the previous step, it will be able to decide whether the first output should be turned on based on the state of the first input.

因为在前面的步骤中，已经知道哪个输入端导通或截止，那么，基于第一个输入端的状态，可以决定第一个输出是否应被触发。

[9] Based on the example in Step 2 it would now turn on the first output because the first input was on and your program said to turn on the first output when this condition is true.

基于第二步中的例子，因为第一个输入端被触发，而且程序要求，当条件为真时，第一个输出被触发,所以,第一个输出就应当被触发。

[10] Since technology for motion control of electric drives became available, the use of programmable logic controllers (PLCs) with power electronics in electric machines applications has been introduced in the manufacturing automation.

由于电气传动运动控制技术的应用,使得在制造业自动化领域中引进了电动机应用中的电力电子可编程序逻辑控制技术。

[11] But most importantly, damage to machine and personnel could result from improper maintenance management of your company's PLCs.

但最重要的是,贵公司的 PLC 维护管理不当可能造成机器损坏和人员损伤。

[12] Just as a doctor asks the patient questions to figure out what is wrong, a maintenance technician asks the PLC questions to troubleshoot the machine.

正如医生询问病人问题,以便找出是什么病因一样,维修技术人员检查 PLC 设备以便找出机器存在的问题。

[13] It is common to only learn about a PLC once the machine is down and the clock is ticking at a thousand dollars an hour, or more.

通常情况,只有当 PLC 故障才会弄清楚它。维修费用为每小时 1 千美元,或者更高。

[14] Also with all the baby boomers (our core knowledge base in the industry) about to retire, it is not smart management to place all your eggs in one basket.

同时,随着所有掌握企业核心技术人员的退休,我们明白,把所有的鸡蛋放在一个篮子里的方法并不是明智的管理方法。

[15] Some other good ideas to get more out of your PLC training investment would be to get hands on training using the actual PLC programs and software the maintenance technician will be working with in the facility.

对于你的 PLC 培训投入其他一些好的建议是应用实际的 PLC 程序和设施中将要使用的技术支持软件进行培训。

[16] During the three decades following their introduction, PLCs have evolved to incorporate analog I/O, communication over networks, and new programming standards.

在产生后的 30 年里,PLC 已经形成了模拟 I/O 的连接、网络通信和新的程序标准。

[17] The applications that fall within the 20 percent are built by engineers who require higher loop rates, advanced control algorithms, more analogy capabilities, and better integration with the enterprise network.

这些应用是由 20% 的工程师完成的,这些工程师要求程序循环使用率更高,控制算法更先进,模拟能力更强,以及企业网络的兼容性更好。

[18] So the "20 percenters" either lived without functionality not easily accomplished with a PLC or cobbled together a system that included a PLC for the control portion of the code and a PC for the more advanced functionality.

因此这 20% 用一个功能性的 PLC,或者用包含一个 PLC 作为代码控制部分和一个完成更先进功能的 PC 形成的组装系统都是不容易实现的。

[19] With no clear PC or PLC solution, engineers with complex applications worked close-

ly with control vendors to develop new products.

没有 PC 或 PLC 清楚的解决方案，从事复杂应用工作的工程师们和控制供应商紧密联系共同开发新的产品。

[20] This enables vendors to develop more powerful software with the flexibility and usability of PC-based control systems that can run on real-time operating systems for reliability.

这使得供应商能开发出基于 PC 控制的灵活性和实用性更强大的软件，并且能运行在实时操作系统中满足可靠性要求。

Chapter 9 Single Chip Microcomputer Control Technology

9.1 Foundation

The single-chip microcomputer is the culmination of both the development of the digital computer and the integrated circuit, arguably the two most significant inventions of the 20th century.[1]

These two types of architecture are found in single-chip microcomputers. Some employ the split program/data memory of the Harvard architecture, shown in Fig. 9-1, others follow the philosophy, widely adopted for general-purpose computers and microprocessors, of making no logical distinction between program and data memory as in the Princeton architecture, shown in Fig. 9-2.

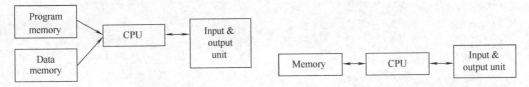

Fig. 9-1 A Harvard Type   Fig. 9-2 A Conventional Princeton Computer

In general terms a single-chip microcomputer is characterized by the incorporation of all the units of a computer into a single device, as shown in Fig. 9-3.

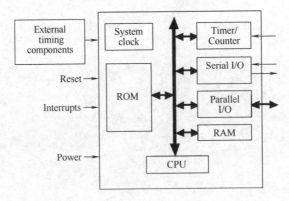

Fig. 9-3 Principal Features of a Microcomputer

9.1.1 Read Only Memory-ROM

ROM is usually for the permanent, non-volatile storage of an applications program. Many microcomputers and micro controllers are intended for high-volume applications and hence the economical manufacture of the devices requires that the contents of the program memory be committed permanently during the manufacture of the chips. Clearly, this implies a rigorous approach to ROM code development since changes cannot be made after manufacture. This development process may involve emulation using a sophisticated development system with a hardware emulation capability as well as the use of powerful software tools.

Some manufacturers provide additional ROM options by including in their range devices with (or intended for use with) user programmable memory. The simplest of these is usually a device which can operate in a microprocessor mode by using some of the input/output lines as an address and data bus for accessing external memory. This type of device can behave functionally as the single-chip microcomputer from which it is derived albeit with restricted I/O and a modified external circuit. The use of these ROMless devices is common even in production circuits where the volume does not justify the development costs of custom on-chip ROM.[2] There is a significant saving in I/O and other chips compared to a conventional microprocessor based circuit. More exact replacements for ROM devices can be obtained in the form of variants with 'piggy-back' EPROM (Erasable Programmable ROM) sockets or devices with EPROM instead of ROM.[3] These devices are naturally more expensive than the equivalent ROM device, but do provide complete circuit equivalents. EPROM based devices are also extremely attractive for low-volume applications where they provide the advantages of a single-chip device, in terms of on-chip I/O, etc., with the convenience of flexible user programmability.

9.1.2 Random Access Memory-RAM

RAM is for the storage of working variables and data used during program execution. The size of this memory varies with device type but it has the same characteristic width (4, 8, 16 bits, etc.) as the processor. Special function registers, such as a stack pointer or timer register are often logically incorporated into the RAM area.[4] It is also common in Harvard type microcomputers to treat the RAM area as a collection of registers; it is unnecessary to make a distinction between RAM and processor register as is done in the case of a microprocessor system since RAM and registers are not usually physically separated in a microcomputer.

9.1.3 Central Processing Unit

The CPU is much like that of any microprocessor. Many applications of microcomputers and single-chip microcomputers involve the handling of binary-coded decimal (BCD) data (for numerical displays, for example), hence it is common to find that the CPU is well adapted to handling this type of data. It is also common to find good facilities for testing, setting and resetting individual bits of memory or I/O since many controller applications involve the turning on and off of single output lines or the reading of a single line. These lines are readily interfaced to two-state devices such as switches, thermostats, solid-state relays, valves, motors, etc.

9.1.4 Parallel Input/Output

Parallel input and output schemes vary somewhat in different microcomputers; in most a mechanism is provided to at least allow some flexibility of choosing which pins are outputs and which are inputs. This may apply to all or some of the ports. Some I/O lines are suitable for direct interfacing to, for example, fluorescent displays, or can provide sufficient current to make interfacing to other components straightforward. Some devices allow an I/O port to be configured as a system bus to allow off-chip memory and I/O expansion. This facility is potentially useful as a product range develops, since successive enhancements may become too big for on-chip memory and it is undesirable not to build on the existing software base.

9.1.5 Serial Input/Output

Serial communication with terminal devices is a common means of providing a link using a small number of lines. This sort of communication can also be exploited for interfacing special function chips or linking several microcomputers together. Both the common asynchronous and synchronous communication schemes require protocols that provide framing (start and stop) information. This can be implemented as a hardware facility or U (S) ART [Universal (synchronous) asynchronous receiver/transmitter] relieving the processor (and the applications programmer) of this low-level, time-consuming, detail. It is merely necessary to select a baud-rate and possibly other options (number of stop bits, parity, etc.) and load (or read from) the serial transmitter (receiver) buffer. [5] Serialization of the data in the appropriate format is then handled by the hardware circuit.

9.1.6 Timer/Counter Facilities

Many applications of single-chip microcomputers require accurate evaluation of elapsed real time. This can be determined by careful assessment of the execution time of each branch in a program but this rapidly becomes inefficient for all but the simplest programs. The preferred approach is to use a timer circuit that can independently count precise time increments and generate an interrupt after a preset time has elapsed. This type of timer is usually arranged to be preloadable with the required count. The timer then decrements this value producing an interrupt or setting a flag when the counter reaches zero. Better timers then have the ability to automatically reload the initial count value. This relieves the programmer of the responsibility of reloading the counter and assessing the elapsed time before the timer is restarted, which otherwise would be necessary if continuous precisely timed interrupts were required (as in a clock, for example). [6] Sometimes associated with a timer is an event counter. With this facility there is usually a special input pin, that can drive the counter directly.

9.1.7 Timing Components

The clock circuitry of most microcomputers requires only simple timing components. If maximum performance is required, a crystal must be used to ensure the maximum clock frequency is approached but not exceeded. Many clock circuits also work with a resistor and capacitor as low-cost timing components or can be driven from an external source. This latter arrangement is useful if external synchronization of the microcomputer is required.

9.2 A Single-Chip Microcomputer Integrated Circuit

A single-chip microcomputer integrated circuit comprises a processor core which exchanges data with at least one data processing and/or storage device. The integrated circuit comprises a mask-programmed read only memory containing a generic program such as a test program which can be executed by single-chip microcomputer. The generic program includes a basic function for writing data into the data processing and/or storage device or devices. The write function is used to load a downloading program. Because a downloading program is not permanently stored in the read only memory, the single-chip microcomputer can be tested independently of the application program, and remains standard with regard to the type of memory component with which it can be used in a system. [7]

To be more precise, the invention concerns a single-chip microcomputer integrated circuit. A single-chip microcomputer is usually a VLSI (Very Large Scale Integration) integrated circuit containing all or most of the components of a "computer". Its function is not predefined but depends on the program that it executes. [8]

A single-chip microcomputer necessarily comprises a processor core including a command sequencer (which is a device distributing various control signals to the other componentsaccording to the instructions of a program), an arithmetic and logic unit (for processing the data) and registers (which are specialized memory units).

The other components of the "computer" can be either internal or external to the single-chip microcomputer, however. [9] In other words, the other components are integrated into either the single-chip microcomputer or auxiliary circuits.

These other components of the "computer" are data processing and/or storage devices, for example, read only or random access memory containing the program to be executed, clocks and interfaces (serial or parallel).

As a general rule, a system based on a single-chip microcomputer therefore comprises a microchip containing the single-chip microcomputer, and a plurality of microchips containing the external data processing and/or storage devices which are not integrated into the single-chip microcomputer. A single-chip microcomputer-based system of this kind comprises, for example, one or more printed circuit boards on which the single-chip microcomputer and the other components are mounted.

It is the application program, i.e. the program which is executed by the single-chip microcomputer, which determines the overall operation of the single-chip microcomputer system. Each application program is therefore specific to a separate application.

In most current applications the application program is too large to be held in the single-chip microcomputer and is therefore stored in a memory external to the single-chip microcomputer. [10] This program memory, which has only to be read, not written, is generally a reprogrammable read only memory (REPROM).

After the application program has been programmed in memory and then started in order to be executed by the single-chip microcomputer, the single-chip microcomputer system may not function as expected.

In the least unfavorable situation this is a minor dysfunction of the system and the single-chip microcomputer is still able to dialog with a test station via a serial or parallel interface. This test station is then able to determine the nature of the problem and indicate precisely the type of correction (software and/or physical) to be applied to the system for it to operate correctly.

Unfortunately, most dysfunctions of a single-chip microcomputer-based system result in a total system lock-up, preventing any dialog with a test station. It is then impossible to determine the type of fault, i. e. whether it is a physical fault (in the single-chip microcomputer itself, in an external read only memory, in a peripheral device, on a bus, etc.) or a software fault (i. e. an error in the application program). [11] The troubleshooting technique usually employed in these cases of total lock-up is based on the use of sophisticated test devices requiring the application of probes to the pins of the various integrated circuits of the single-chip microcomputer-based system under test.

There are various problems associated with the use of such test devices for troubleshooting a single-chip microcomputer-based system. The probes used in these test devices are very fragile, difficult to apply because of the small size of the circuit and their close packing, and may not make good contact with the circuit. [12]

Also, because of their high cost, these test devices are not mass produced. Consequently, faulty microcontroller-based systems cannot be repaired immediately, wherever they happen to be located at the time, but must first be returned to a place where a test device is available. Troubleshooting a single-chip microcomputer-based system in this way is time-consuming, irksome and costly.

To avoid the need for direct action on the single-chip microcomputer-based system each time the application program executed by the single-chip microcomputer of the system is changed, it is standard practice to use a downloadable read only memory to store the application program, a loading program being written into a mask-programmed read only memory of the single-chip microcomputer. The mask-programmed read only memory of the single-chip microcomputer is integrated into the read only memory of the single-chip microcomputer and programmed once and for all during manufacture of the single-chip microcomputer. [13]

To change the application program the read only memory of the single-chip microcomputer is reset by running the downloading program. This downloading program can then communicate with a workstation connected to the read only memory of the single-chip microcomputer by an appropriate transmission line, this workstation containing the new application program to be written into the single-chip microcomputer. The downloading program receives the new application program and loads it into a read only memory external to the single-chip microcomputer. [14]

Although this solution avoids the need for direct action on the single-chip microcomputer-based system (which would entail removing from the system the reprogrammable read only memories containing the application program, writing into these memories the new application program using an appropriate programming device and then replacing them in the system), it nevertheless has a major drawback, namely specialization of the single-chip microcomputer during manufacture.

Each type of reprogrammable memory is associated with a different downloading program because the programming parameters (voltage to be applied, duration for which the voltage is to be applied, etc.) vary with the technology employed. The downloading program is written once for all into the mask-programmed internal memory of the single-chip microcomputer and the latter is therefore restricted to using memory components of the type for which this downloading program was written. In other words, the single-chip microcomputer is not a standard component and this increases its cost of manufacture.

One object of the invention is to overcome these various drawbacks of the prior art. To be more precise, an object of the invention is to provide a single-chip microcomputer circuit which can verify quickly, simply, reliably and at low cost the operation of a system based on the single-chip microcomputer.

Another object of the invention is to provide a single-chip microcomputer integrated circuit which can accurately locate the defective component or components of a system using the single-chip microcomputer in the event of dysfunction of the system. [15]

A further object of the invention is to provide a single-chip microcomputer integrated circuit which avoids the need for direct action on the single-chip microcomputer-based system to change the application program, whilst remaining standard as regards the type of memory component with which it can be used in a system.

9.3 Digital Signal Processors

A Digital Signal Processor is a super-fast chip computer, which has been optimized for the detection, processing and generation of real world signals such as voice, video, music, etc, in real time. [16]

It is usually implemented in a single chip, or nowadays just of an IC, about 0.5 cm^2 to 4 cm^2, costing between $ 3 and $ 300 and residing in every mobile phone, MP3 player, computer and most cars on our planet.

In contrast, a microprocessor (a term with which we are all probably familiar) is traditionally a much less powerful computer that performs the mundane "behind the scenes" tasks, often controlling other devices—e.g., keyboard entry, central heating, washing machine cycles, etc. Having said this, a Pentium III chip normally be classed as a microprocessor rather than a DSP, yct, running at speeds in excess of 1 GHz, this can hardly be described as "much less powerful".

Today there is in fact a blurring between the roles of microprocessor and digital signal processor—perhaps synergy is a better word. [17]

Many so-called DSPs now have micro-processor, functionality on board. Many so-called micros now have DSP functionality on board. Distinguishing between DSP and micro by application is perhaps not the best way forward after all. In fact, the main difference lies deep inside the devices themselves, in the internal chip architecture, with DSP devices particularly optimized for high-speed, high-accuracy multiplication.

To "digital signal process" is to manipulate signals that have either originated in, or are to be exported to, the real world, where those signals are represented as digits (numbers). This means that an integral part of DSP applications is the conversion of the real world signals-analog voice, music, video, engine speed, ground vibration—to numerical values for processing by the DSP—a process termed analog to digital conversion. Going in the other direction—the conversion of numerical values generated within the DSP to real world signal—a process termed digital to analog conversion is also involved.

Fundamental to DSP is the fact that any real world signal, e.g. music, can be accurately represented by samples of the signal taken at periodic intervals. These samples can then be converted into numbers and these numbers are expressed in binary form. We are now in the world of computing and software where manipulation of binary numbers is bread and butter stuff. [18]

If we simply scale all of the numbers representing our samples by an equal amount, and then convert the numbers back to voltage samples to drive our headphones or speakers, we have implemented a DSP volume control. If on the other hand, we scale the numbers by a different amount each second, we have the beginnings of a simple audio FX processor.

In DSP terms, these are very simple tasks, and do not even begin to harness the processing power available. An example of a more challenging application is a graphic equalizer. Essentially this allows you to change the volume to components of the music source depending on their frequency. [19] To implement a graphic equalizer requires the separation of the signal into frequency bands using filters, scaling each filtered segment independently, and then adding the segments back together for sending to the output.

Digital filtering is child's play for a DSP, but actually involves a lot of computation. Consecutive samples must be stored, each scaled independently and then summed to perform even the most simple of filter functions. [20] However, the DSP and its instruction set are optimized for tasks such as this (hence the distinction between a DSP and a microprocessor), and this function can indeed be implemented using just three lines of code on most DSP devices.

New Words and Expressions

1. culmination   *n.* 顶点，极点
2. split   *adj.* 分离的
3. philosophy   *n.* 原理，原则
4. incorporation   *n.* 合并，结合
5. volatile   *adj.* 易变的
6. commit   *v.* 保证
7. emulation   *n.* 竞争
8. albeit   *conj.* 虽然
9. custom   *adj.* 定制
10. variant   *adj.* 不同的，替换的

11. piggy-back   *adj.* 背负式的
12. erasable   *adj.* 可擦除的
13. socket   *n.* 插座
14. thermostat   *n.* 恒温器
15. protocol   *n.* 协议
16. time-consuming   耗时的
17. baud   *n.* 波特
18. elapse   *v.* 经过
19. evaluation   *n.* 估计
20. preset   *adj.* 事先调整的
21. preloadable   *adj.* 预载的
22. decrement   *n.* 减少量
23. auxiliary   *adj.* 辅助的
24. drawback   *n.* 缺点，障碍
25. dysfunction   *n.* 官能不良，官能障碍
26. interface   *n.* 接口，界面，连接体
27. irksome   *adj.* 令人厌烦的，令人恨恼的
28. namely   *adv.* 即，那就是
29. peripheral   *adj.* 次要的，外围的   *n.* 外围设备
30. sequencer   *n.* 程序装置，定序器
31. specialization   *n.* 特别化，专门化
32. troubleshooting   *n.* 故障诊断

Notes

［1］The single-chip microcomputer is the culmination of both the development of the digital computer and the integrated circuit, arguably the two most significant inventions of the 20th century.
单片机是数字计算机和集成电路发展中的一个顶峰，而数字计算机和集成电路可以说是 20 世纪两项最有意义的发明。

［2］The use of these ROMless devices is common even in production circuits where the volume does not justify the development costs of custom on-chip ROM.
这些无 ROM 设备的应用即使在生产电路中也是常见的，生产电路的数量不能改变定制的片内 ROM 的开发。

［3］More exact replacements for ROM devices can be obtained in the form of variants with 'piggy-back' EPROM (Erasable Programmable ROM) sockets or devices with EPROM instead of ROM.
ROM 设备更精确的替代可以通过不同形式的带有背负式 EPROM（可擦写编程 ROM）插座或由 EPROM 取代 ROM 的设备而获得。

［4］Special function registers, such as a stack pointer or timer register are often logically incorporated into the RAM area.
特殊功能寄存器，如堆栈指针或时间寄存器通常在逻辑上合并到 RAM 区域。

［5］It is merely necessary to select a baud-rate and possibly other options (number of stop bits, parity, etc.) and load (or read from) the serial transmitter (receiver) buffer.
需要选择一个波特率和其他可能的选项（停止位的数目、奇偶检验等），以及装载（或读取）串行发送器（或接收器）的缓冲区。

［6］This relieves the programmer of the responsibility of reloading the counter and assessing the elapsed time before the timer is restarted, which otherwise would be necessary if continuous precisely timed interrupts were required (as in a clock, for example).
这使得程序员从重新加载计数器的数据和设置定时器再启动前的经过时间的工作中解脱出来，否则的话，对于有连续精确定时中断要求（的项目），这些工作是必须要做的（例

如，在一个时钟周期内）。

[7] Because a downloading program is not permanently stored in the read only memory, the single-chip microcomputer can be tested independently of the application program, and remains standard with regard to the type of memory component with which it can be used in a system.

因为装载程序并非永久地存储在只读存储器中，所以可对单片机进行测试，而与应用程序无关，并保持系统中能用的存储器元件为标准类型。

[8] Its function is not predefined but depends on the program that it executes.

它的功能不是预先确定的，而是取决于它所执行的程序。

[9] The other components of the "computer" can be either internal or external to the single-chip microcomputer, however.

然而对单片机而言，"计算机"的其他部件可以是内部的或是外部的。

[10] In most current applications the application program is too large to be held in the single-chip microcomputer and is therefore stored in a memory external to the single-chip microcomputer.

在多数的实际应用中，由于应用程序太大，单片机无法存储，因此就存储在单片机的外部存储器中。

[11] It is then impossible to determine the type of fault, i.e. whether it is a physical fault (in the single-chip microcomputer itself, in an external read only memory, in a peripheral device, on a bus, etc) or a software fault (i.e. an error in the application program).

这样就不能确定错误类型是硬件错误（单片机本身、外部只读存储器、外围设备、总线等）还是软件错误（应用程序的错误）。

[12] The probes used in these test devices are very fragile, difficult to apply because of the small size of the circuit and their close packing, and may not make good contact with the circuit.

这些测试设备中使用的探针非常脆弱，因为应用的电路尺寸小，包装紧密并且可能和电路接触不好，所以难以适用。

[13] The mask-programmed read only memory of the single-chip microcomputer is integrated into the read only memory of the single-chip microcomputer and programmed once and for all during manufacture of the single-chip microcomputer.

单片机的掩膜编程只读内存集成到单片机的只读存储器内并且程序可以一劳永逸地固化在单片机内。

[14] The downloading program receives the new application program and loads it into a read only memory external to the single-chip microcomputer.

下载程序接收到新的应用程序并且加载到一个单片机外部的只读存储器内。

[15] Another object of the invention is to provide a single-chip microcomputer integrated circuit which can accurately locate the defective component or components of a system using the single-chip microcomputer in the event of dysfunction of the system.

另一个发明的目标是提供一个单片机集成电路，它可以在系统发生功能障碍时准确定位系统有缺陷的组件。

[16] A Digital Signal Processor is a super-fast chip computer, which has been optimized for

Chapter 9 Single Chip Microcomputer Control Technology

the detection, processing and generation of real world signals such as voice, video, music, etc, in real time.

数字信号处理器是一种超高速的计算机芯片,它对真实世界的信号如语音、视频、音乐等已经进行了检测优化和处理。

[17] Today there is in fact a blurring between the roles of microprocessor and digital signal processor—perhaps synergy is a better word.

当前,实际上微处理器和数字信号处理器两者的任务界限已经变得模糊了——协同可能是一个描述两者的关系的较好的词。

[18] We are now in the world of computing and software where manipulation of binary numbers is bread and butter stuff.

我们现在生活在计算机和软件的世界中,在这个世界中,二进制数字的处理就好比是面包和黄油,不可缺少。

[19] Essentially this allows you to change the volume to components of the music source depending on their frequency.

实质上,这允许你根据乐源组件的频率来改变音量。

[20] Digital filtering is child's play for a DSP, but actually involves a lot of computation. Consecutive samples must be stored, each scaled independently and then summed to perform even the most simple of filter functions.

数字滤波对 DSP 来说是非常容易实现的事,但实际上涉及大量的计算。即使是实现最简单的滤波器功能也要进行连续采样存储,每个数据独立地测量,然后进行求和。

Chapter 10 Computer Networking Basics

10.1 Foundation

Suppose you want to build a computer network, one that has the potential to grow to global proportions and to support applications as diverse as teleconferencing, video-on-demand, electronic commerce, distributed computing, and digital libraries. What available technologies would serve as the underlying building blocks, and what kind of software architecture would you design to integrate these building blocks into an effective communication service? Answering this question is the overriding goal of this book—to describe the available building materials and then to show how they can be used to construct a network from the ground up. [1]

Before we can understand how to design a computer network, we should first agree on exactly what a computer network is. At one time, the term *network* meant the set of serial lines used to attach dumb terminals to mainframe computers; To some, the term implies the voice telephone network; To others, the only interesting network is the cable network used to disseminate video signals. The main thing these networks have in common is that they are specialized to handle one particular kind of data (keystrokes, voice, or video) and they typically connect to special-purpose devices (terminals, hand receivers, and television sets).

What distinguishes a computer network from these other types of networks? Probably the most important characteristic of a computer network is its generality. Computer networks are built primarily from general-purpose programmable hardware, and they are not optimized for a particular application like making phone calls or delivering television signals. [2] Instead, they are able to carry many different types of data, and they support a wide, and ever-growing, range of applications. This chapter looks at some typical applications of computer networks and discusses the requirements that a network designer who wishes to support such applications must be aware of.

Once we understand the requirements, how do we proceed? Fortunately, we will not be building the first network. Others, most notably the community of researchers responsible for the Internet, have gone before us. We will use the wealth of experience generated from the Internet to guide our design. This experience is embodied in a network architecture that identifies the available hardware and software components and shows how they can be arranged to form a complete network system. [3]

10.2 Applications

Most people know the Internet through its applications: the World Wide Web, email, streaming audio and video, chat rooms, and music (file) sharing. The Web, for example, presents an intuitively simple interface. Users view pages full of textual and graphical objects, click on objects that they want to learn more about, and a corresponding new page appears. Most people are also aware that just under the covers, each selectable object on a page is bound to an identifier for the next page to be viewed. This identifier, called a Uniform Resource Locator (URL), is used to provide a way of identifying all the possible pages that can be viewed from your web browser.

Another widespread application of the Internet is the delivery of "streaming" audio and video. While an entire video file could first be fetched from a remote machine and then played on the local machine, similar to the process of downloading and displaying a web page, this would entail waiting for the last second of the video file to be delivered before starting to look at it. Streamingvideo implies that the sender and the receiver are, respectively, the source and the sink for the video stream. [4]

That is, the source generates a video stream (perhaps using a video capture card), sends it across the Internet in messages, and the sink displays the stream as it arrives.

There are a variety of different classes of video applications. One class of video application is video-on-demand, which reads a preexisting movie from disk and transmits it over the network. Another kind of application is videoconferencing, which is in some ways the more challenging and interesting case for networking people because it has very tight timing constraints. Just as when using the telephone, the interactions among the participants must be timely. When a person at one end gestures, then that action must be displayed at the other end as quickly as possible. Too much delay makes the system unusable. Contrast this with video-on-demand where, if it takes several seconds from the time the user starts the video until the first image is displayed, the service is still deemed satisfactory. [5] Also, interactive video usually implies that video is flowing in both directions, while a video-on-demand application is most likely sending video in only one direction.

10.3 Requirements

We have just established an ambitious goal for ourselves: to understand how to build a computer network from the ground up. Our approach to accomplishing this goal will be to start from first principles, and then ask the kinds of questions we would naturally ask if building an actual network. At each step, we will use today's protocols to illustrate various design choices available to us, but we will not accept these existing artifacts as gospel. Instead, we will be asking (and answering) the question of why networks are designed the way they are. While it is tempting to settle for just understanding the way it's done today, it is important to recognize the underlying concepts because networks are constantly changing as the technology evolves and new applications are invented. [6] It is our experience that once you understand the fundamental ideas, any protocol that you are confronted with will be relatively easy to digest.

The first step is to identify the set of constraints and requirements that influence network design. Before getting started, however, it is important to understand that the expectations you have of a network depend on your perspective.

1) An *application programmer* would list the services that his application needs, for example, a guarantee that each message the application sends will be delivered without error within a certain amount of time.

2) A *network designer* would list the properties of a cost-effective design, for example, that network resources are efficiently utilized and fairly allocated to different users.

3) A *network provider* would list the characteristics of a system that is easy to administer and manage, for example, in which faults can be easily isolated and where it is easy to account for usage.

This section attempts to distill these different perspectives into a high-level introduction to the major considerations that drive network design, and in doing so, identifies the challenges addressed throughout the rest of this book.

10.4　Links, Nodes and Clouds

Network connectivity occurs at many different levels. At the lowest level, a network can consist of two or more computers directly connected by some physical medium, such as a coaxial cable or an optical fiber. We call such a physical medium a link, and we often refer to the computers it connects as nodes. (Sometimes a node is a more specialized piece of hardware rather than a computer, but we overlook that distinction for the purposes of this discussion.) As illustrated in Fig. 10-1, physical links are sometimes limited to a pair of nodes (such a link is said to be point-to-point), while in other cases, more than two nodes may share a single physical link (such a link is said to be multiple-access). Whether a given link supports point-to-point or multiple-access connectivity depends on how the node is attached to the link. It is also the case that multiple-access links are often limited in size, in terms of both the geographical distance they can cover and the number of nodes they can connect.

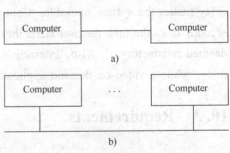

Fig. 10-1　Direct Links
a) Point-to-Point　b) Multiple-Access

If computer networks were limited to situations in which all nodes are directly connected to each other over a common physical medium, then networks would either be very limited in the number of computers they could connect, or the number of wires coming out of the back of each node would quickly become both unmanageable and very expensive. Fortunately, connectivity between two nodes does not necessarily imply a direct physical connection between them—indirect connectivity may be achieved among a set of cooperating nodes.[7] Consider the following two examples of how a collection of computers can be indirectly connected.

Fig. 10-2 shows a set of nodes, each of which is attached to one or more point-to-point links. Those nodes that are attached to at least two links run software that forwards data received on one link out on another. If organized in a systematic way, these forwarding nodes form a switchednetwork. There are numerous types of switched network, of which the two most common are circuit-switched and packet-switched.[8] The former is most notably employed by the telephone system, while the latter is used for the overwhelming majority of computer networks and will be the focus of this book. This important feature of packet-switched networks is that the nodes in such a network send discrete blocks of data to each other. Think of these blocks of data as corresponding to some piece of application data such as a file, a piece of email, or an image. We call each block of data either a packet or a message, and for now we use these terms interchangeably, we discuss the reason they are not always the same in next section.

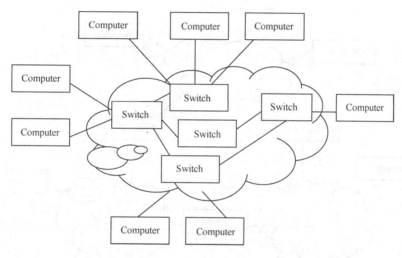

Fig. 10-2 Switched Network

Packet-switched networks typically use a strategy called store-and-forward. As the name suggests, each node in a store-and-forward network first receives a complete packet over some link, stores the packet in its internal memory, and then forwards the complete packet to the next node. In contrast, a circuit-switched network first establishes a dedicated circuit across a sequence of links and then allows the source node to send a stream of bits across this circuit to a destination node.[9] The major reason for using packet switching rather than circuit switching in a computer network is efficiency, discussed in the next subsection.

The cloud in Fig. 10-2 distinguishes between the nodes on the inside that implement the network (they are commonly called switches, and their primary function is to store and forward packets) and the nodes on the outside of the cloud that use the network (they are commonly called hosts, and they support users and run application programs). Also note that the cloud in Fig. 10-2 is one of the most important icons of computer networking. In general, we use a cloud to denote any type of network, whether it is a single point-to-point link, a multiple-access link, or a switched network. Thus, whenever you see a cloud used in a figure, you can think of it as a placeholder for any of the networking technologies covered in this book.

A second way in which a set of computers can be indirectly connected is shown in Fig. 10-3.

In this situation, a set of independent networks (clouds) are interconnected to form an internetwork, or Internet for short. We adopt the Internet's convention of referring to a generic internetwork of networks as a lowercase *i* Internet, and the currently operational TCP/IP Internet as the capital *I* Internet. [10] A node that is connected to two or more networks is commonly called a router or gateway, and it plays much the same role as a switch—it forwards messages from one network to another. Note that an Internet can itself be viewed as another kind of network, which means that an Internet can be built from an interconnection of Internets. Thus, we can recursively build arbitrarily large networks by interconnecting clouds to form larger clouds.

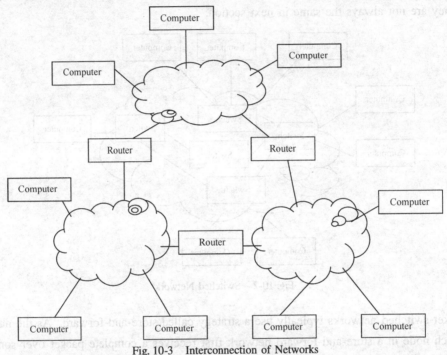

Fig. 10-3   Interconnection of Networks

Just because a set of hosts are directly or indirectly connected to each other does not mean that we have succeeded in providing host-to-host connectivity. The final requirement is that each node must be able to state which of the other nodes on the network it wants to communicate with. This is done by assigning an address to each node. An address is a byte string that identifies a node; that is, the network can use a node's address to distinguish it from the other nodes connected to the network. [11] When a source node wants the network to deliver a message to a certain destination node, it specifies the address of the destination node. If the sending and receiving nodes are not directly connected, then the switches and routers of the network use this address to decide how to forward the message toward the destination. The process of determining systematically how to forward messages toward the destination node based on its address is called routing.

10.5   Network Architecture

In case you hadn't noticed, the previous section established a pretty substantial set of requirements for network design—a computer network must provide general, cost-effective, fair, and robust connectivity among a large number of computers. As if this weren't enough, networks do not remain fixed at any single point in time, but must evolve to accommodate changes in both the underlying technologies upon which they are based as well as changes in the demands placed on them by application programs. [12] Designing a network to meet these requirements is not a small task.

To help deal with this complexity, network designers have developed general blueprints-usually called network architectures—that guide the design and implementation of networks. This section defines more carefully what we mean by a network architecture by introducing the central ideas that are common to all network architectures. It also introduces two of the most widely referenced architectures—the OSI architecture and the Internet architecture.

When a system gets complex, the system designer introduces another level of abstraction. The idea of an abstraction is to define a unifying model that can capture some important aspect of the system, encapsulate this model in an object that provides an interface that can be manipulated by other components of the system, and hide the details of how the object is implemented from the users of the object. The challenge is to identify abstractions that simultaneously provide a service that proves useful in a large number of situations and that can be efficiently implement in the underlying system. This is exactly what we were doing when we introduced the idea of a channel in the previous section: We were providing an abstraction for applications that hides the complexity of the network from application writers. [13]

Abstractions naturally lead to layering, especially in network systems. The general idea is that you start with the services offered by the underlying hardware, and then add a sequence of layers, each providing a higher (more abstract) level of service. The services provided at the high layers are implemented in terms of the services provided by the low layers. Drawing on the discussion of requirements given in the previous section, for example, we might imagine a simple network as having two layers of abstraction sandwiched between the application program and the underlying hardware, as illustrated in Fig. 10-4. The layer immediately above the hardware in this case might provide host-to-host connectivity, abstracting away the fact that there may be an arbitrarily complex network topology between any two hosts. [14] The next layer up builds on the available host-to-host communication service and provides support for process-to-process channels, abstracting away the fact that the network occasionally loses messages, for example.

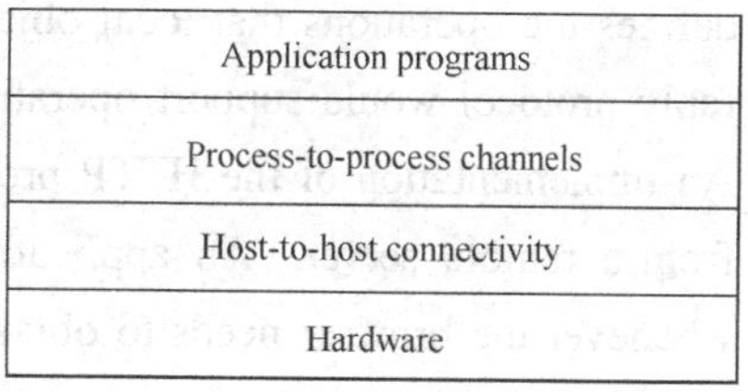

| Application programs |
| Process-to-process channels |
| Host-to-host connectivity |
| Hardware |

Fig. 10-4 Example of a Layered Network System

Layering provides two nice features. First, it decomposes the problem of building a network into more manageable components. Rather than implementing a monolithic piece of software that does everything you will ever want, you can implement several layers, each of which solves one part of the problem. [15] Second, it provides a more modular design. If you decide that you want to add some new service, you may only need to modify the functionality at one layer, reusing the functions provided at all the other layers.

Thinking of a system as a linear sequence of layers is an oversimplification, however. Many times there are multiple abstractions provided at any given level of the system, each providing a different service to the higher layers but building on the same low-level abstractions. One provides a request/reply service and one supports a message stream service. These two channels might be alternative offerings at some level of a multilevel networking system, as illustrated in Fig. 10-5.

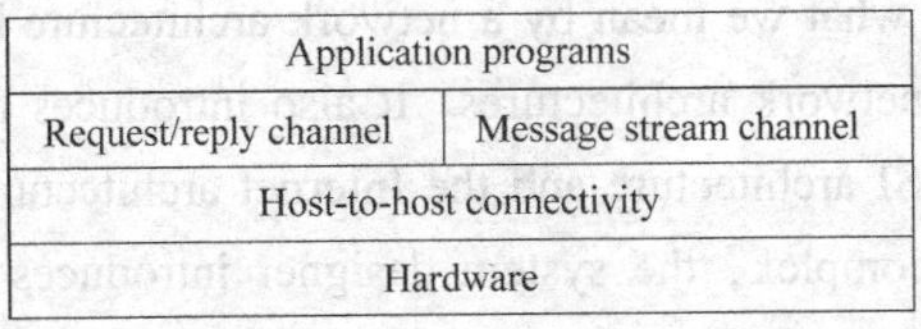

Application programs	
Request/reply channel	Message stream channel
Host-to-host connectivity	
Hardware	

Fig. 10-5 Layered System with Alternative Abstractions Available at a Given Layer

Using this discussion of layering as a foundation, we are now ready to discuss the architecture of a network more precisely. For starters, the abstract objects that make up the layers of a network system are called protocols. That is, a protocol provides a communication service that higher-level objects (such as application processes, or perhaps higher-level protocols) use to exchange messages. For example, we could imagine a network that supports a request/reply protocol and a message stream protocol, corresponding to the request/reply and message stream channels discussed above.

Each protocol defines two different interfaces. First, it defines a service interface to the other objects on the same computer that want to use its communication services. This service interface defines the operations that local objects can perform on the protocol.[16] For example, a request/reply protocol would support operations by which an application can send and receive messages. An implementation of the HTTP protocol could support an operation to fetch a page of hypertext from a remote server. An application such as a web browser would invoke such an operation whenever the browser needs to obtain a new page, for example, when the user clicks on a link in the currently displayed page.

Second, a protocol defines a peer interface to its counterpart (peer) on another machine. This second interface defines the form and meaning of messages exchanged between protocol peers to implement the communication service. This would determine the way in which a request/reply protocol on one machine communicates with its peer on another machine. In the case of HTTP, for example, the protocol specification defines in detail how a "GET" command is formatted, what arguments can be used with the command, and how a web server should respond when it receives such a command.

To summarize, a protocol defines a communication service that it exports locally (the service interface), along with a set of rules governing the messages that the protocol exchanges with its peer(s) to implement this service (the peer interface). This situation is illustrated in Fig. 10-6.

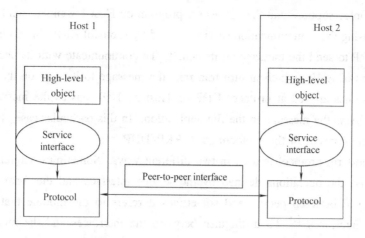

Fig. 10-6 Service and Peer Interfaces

Except at the hardware level where peers directly communicate with each other over a link, peer-to-peer communication is indirect—each protocol communicates with its peer by passing messages to some lower-level protocol, which in turn delivers the message to its peer. In addition, there are potentially multiple protocol at any given level, each providing a different communication service. We therefore represent the suite of protocols that make up a network system with a protocol graph. The nodes of the graph correspond to protocols, and the edges represent a depends on relation. For example, Fig. 10-7 illustrates a protocol graph for the hypothetical layered system we have been discussing—the protocols Request/Reply Protocol (RRP) and Message Stream Protocol (MSP) implement two different types of process-to-process channels, and both depend on Host-to-Host (HHP), which provides a host-to-host connectivity service.

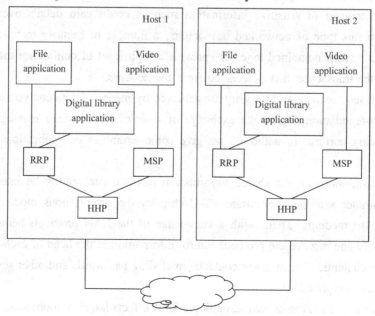

Fig. 10-7 Example of a Protocol Graph

In this example, suppose that the file access program on Host 1 wants to send a message to its peer on Host 2 using the communication service offered by protocol RRP. In this case, the file application asks RRP to send the message on its behalf. To communicate with its peer, RRP then invokes the services of HHP, which in turn transmits the message to its peer on the other machine. Once the message has arrived at protocol HHP on Host 2, HHP passes the message up to RRP, which in turn delivers the message to the file application. In this particular case, the application is said to employ the services of the protocol stack RRP/HHP.

Note that the term protocol is used in two different ways. Sometimes it refers to the abstract interfaces—that is, the operations defined by the service interface and the form and meaning of messages exchanged between peers, and sometimes it refers to the module that actually implements these two interfaces. [17] To distinguish between the interfaces and the module that implements these interfaces, we generally refer to the former as a protocol specification. Specifications are generally expressed using a combination of prose, state transition diagrams, pictures of packet formats, and other abstract notations. It should be the case that a given protocol can be implemented in different ways by different programmers, as long as each adheres to the specification. The challenge is ensuring that two different implementations of the same specification can successfully exchange messages. Two or more protocol modules that do accurately implement a protocol specification are said to inter-operate with each other.

10.6 Network Security

Increasingly people are using networks such as the Internet for on-line banking, shopping and many other applications. The generic term used is electronic commerce or e-commerce and this often involves the transfer of sensitive information such as credit card details over the network. Hence to support this type of networked transaction, a number of security techniques have been developed which, when combined together, provide a high level of confidence that any information relating to the transaction that is received from the network.

As we shall see, secrecy and integrity are achieved by means of data encryption while authentication and non-repudiation require the exchange of a set of (encrypted) messages between the two communicating parties. In addition, we give some examples of applications that use these techniques.

LANs, transmissions on the shared transmission medium can readily be intercepted by any system if an intruder sets the appropriate MAC chip-set the promiscuous mode and records all transmission on the medium. Then, with a knowledge of the LAN protocols being used, the intruder can identify and remove the protocol control information at the head of each message, leaving the message contents. The message contents, including passwords and other sensitive information, can then be interpreted.

This is known as listening or eavesdropping and its effects are all too obvious. In addition and perhaps more sinister, an intruder can use a recorded message sequence to generate a new sequence. This is known as masquerading and again the effects are all too apparent.[18] Therefore, encryption should be applied to all data transfers that involve a network. In the context of the TCP/IP reference model, the most appropriate layer to perform such operations is the application layer.

Public key systems like RSA are particularly useful for non-repudiation, that is, providing that a person sent an electronic document. With a paper document, normally a person adds his or her signature at the end of the document—sometimes with the name and signature of a witness—and, should it be necessary, this is then used to verify that the person whose signature is on the document sent it.

One solution is to exploit the dual property of public key systems, namely that not only is a receiver able to decipher all messages it receives using its own private key, but any receiver can also decipher a message encrypted with the sender's private key, using the sender's public key. Although this is an elegant solution, it has a number of limitations.

One solution is to compute a much shorter version of the message based on the message contents, similar in principle to the computation of a CRC. The shorter version is called the message digest (MD) and the computation function that is used to compute it the hash function.[19] The MD is first computed using the chosen hash function. This is then encrypted using the sender's private key. The encrypted MD is then sent together with the plain-text message. At the receiver, the encrypted MD is decry-ted using the sender's public key. The MD of the received plain-text message is also computed and, if this is the same as the decry-ted MD, this is taken as proof that no one has tampered with the message and the sender whose public key was used to decry-pt the MD did in fact send the message.

In general, authentication is required when a client wishes to access some information or service from a networked server. Before the client is allowed access to the server, he or she must first prove to the server that they are a registered user. Once authenticated the user is then allowed access. The authentication process can be carried out using either a public key or a private key scheme.

Security features are located in various parts of a protocol stack. Typically, in fixed-wire networks these start at the network layer where an IETF protocol called IP security is used. There is then a transport layer protocol called Transport Layer Security that is a derivative of a protocol called the Secure Socket Layer. As we have just seen, at the application layer, a security feature is provided for e-mail by a protocol called Pretty Good Privacy. There are also security features associated with the Web.

In general, because of their mode of operation, even tighter levels of security are provided in wireless networks. These include Wired Equivalent Privacy that is used in IEEE802.11 wireless LANs. In addition, Bluetooth has a number of security features WAP2.0 has a full set of security protocols.[20]

When carrying out a transaction over the Web relating to an e-commerce application, since in many instances this involves sending details of a user's payment card, the security of such transfers is vitally important. For example, an eavesdropper with knowledge of Internet protocols could readily intercept the information entered on the order form and, having got the name and other details about the card, proceed to use these to carry out purchases of their own. A second potential pitfall is that the Web site from where the purchase is being made may not in reality have anything for sale and, prior to the agreed delivery date, abscond with the money that it has collected. In addition, in electronic banking (e-banking) and other financial applications, a client may masquerade an another person. Any security scheme, therefore, must be able to counter each of these threats.

New Words and Expressions

1. network   n. 网络，电路
2. potential   n. 潜力，潜在性
3. building blocks   积木
4. cable   n. 电缆，电报
5. keystroke   n. 击键
6. network architecture   网络结构
7. hardware and software   软硬件
8. metric   adj. 十进制的
9. streaming   n. 流
10. video-on-demand   视频点播
11. protocol   n. 协议，协定，礼仪
12. gospel   n. 福音，原则，信条
13. point-to-point   点对点
14. multiple-access   多进程
15. store-and-forward   存储转发
16. host-to-host   主机到主机
17. router   n. 路由器
18. blueprint   n. &v. 蓝图，晒图
19. capture   v. 捕获，占领
20. encapsulate   v. 封装
21. manipulate   v. 操纵，操作
22. simultaneously   adv. 同时地
23. underlying   adj. 底层的，内在的
24. layer   n. 层
25. topology   n. 拓扑
26. process-to-process   进程到进程
27. monolithic   adj. 单片的，独石的
28. modular   adj. 模块化
29. multiple   n. 倍数   adj. 多的，复杂的
30. service interface   应用界面
31. fetch   n. 手法，手腕   v. 取用
32. invoke   v. 呼吁，请求，调用

Notes

[1] Answering this question is the overriding goal of this book—to describe the available building materials and then to show how they can be used to construct a network from the ground up.

本书的首要目标就是回答这个问题——描述必要的构成材料和如何构建一个网络系统。

[2] Computer networks are built primarily from general-purpose programmable hardware, and they are not optimized for a particular application like making phone calls or delivering television signals.

计算机网络主要用于通用编程硬件，它们对于打电话或传送电视信号这样特定的应用没有优化。

[3] This experience is embodied in a network architecture that identifies the available hardware and software components and shows how they can be arranged to form a complete network

system.

这种方法应用在网络结构中,它确定可用的软硬件组件并且表明它们怎样被组合形成一个完整的网络系统。

[4] Streaming video implies that the sender and the receiver are, respectively, the source and the sink for the video stream.

流媒体视频表明发送者和接收者各自独立为视频流的源头和终端。

[5] Contrast this with video-on-demand where, if it takes several seconds from the time the user starts the video until the first image is displayed, the service is still deemed satisfactory.

与视频点播相反,如果从用户开始启动后需要几秒后才开始工作,这仍然被视为是令人满意的。

[6] While it is tempting to settle for just understanding the way it is done today, it is important to recognize the underlying concepts because networks are constantly changing as the technology evolves and new applications are invented.

尽管人们很容易满足于仅仅了解网络的表面情况,但是,认识网络的本质概念是非常重要的,因为网络是随着其包含的技术和新发明应用的变化而不断变化的。

[7] Fortunately, connectivity between two nodes does not necessarily imply a direct physical connection between them——indirect connectivity may be achieved among a set of cooperating nodes.

幸运的是,两个节点之间的连接并不一定意味着它们之间必须有直接的物理连接——它们之间的非直接连接可以通过一组互相合作节点完成。

[8] There are numerous types of switched network, of which the two most common are circuit-switched and packet-switched.

交换网络存在多种形式,其中最常见的两种形式是电路交换和分组交换。

[9] In contrast, a circuit-switched network first establishes a dedicated circuit across a sequence of links and then allows the source node to send a stream of bits across this circuit to a destination node.

相比之下,开关电路网络通过一系列的链接建立了一个专门的电路,通过它允许源节点发送比特流到目标节点。

[10] We adopt the Internet's convention of referring to a generic inter-network of networks as a lowercase *i* Internet, and the currently operational TCP/IP Internet as the capital *I* Internet.

我们采用的网络约定是,通用互联网作为小写 i 网络,当前运行的 TCP／IP 互联网作为大写 I 网络。

[11] An address is a byte string that identifies a node; that is, the network can use a node's address to distinguish it from the other nodes connected to the network.

一个地址是一个字节字符串,它用来标志一个节点;也就是说,网络可以使用一个节点的地址来区别于网络系统中的其他节点。

[12] As if this weren't enough, networks do not remain fixed at any single point in time, but must evolve to accommodate changes in both the underlying technologies upon which they are based as well as changes in the demands placed on them by application programs.

好像这还不够,网络不会保持固定在任何单一时间点,但是,必须适应它们基本内在技术的变化和应用程序发出命令时的转变。

[13] We were providing an abstraction for applications that hides the complexity of the network from application writers.

我们提供应用程序的抽象内容,它隐藏了网络系统的复杂性。

[14] The layer immediately above the hardware in this case might provide host-to-host connectivity, abstracting away the fact that there may be an arbitrarily complex network topology between any two hosts.

在这种情况下,该层立即在硬件上提供主机到主机连接,抽象出来的事实是在任何两个主机之间都可以存在一个任意复杂的网络拓扑结构。

[15] Rather than implementing a monolithic piece of software that does everything you will ever want, you can implement several layers, each of which solves one part of the problem.

而不是要实现软件的每一个细节部分,可以使用几个层,每一层用来解决问题的一个部分。

[16] This service interface defines the operations that local objects can perform on the protocol.

该服务接口定义了本地对象可以按操作协议执行。

[17] Sometimes it refers to the abstract interfaces—that is, the operations defined by the service interface and the form and meaning of messages exchanged between peers, and sometimes it refers to the module that actually implements these two interfaces.

有时,它指的是抽象接口,即是由应用界面定义的操作和相互之间交换信息的形式和含义。有时,它指的就是实际实现这两个接口的模块。

[18] This is known as masquerading and again the effects are all too apparent.

这就是所谓的伪装技术,其效果非常明显。

[19] The shorter version is called the message digest (MD) and the computation function that is used to compute it the hash function.

简要版本被称为信息摘要(MD),用来计算的函数是散列函数。

[20] In addition, Bluetooth has a number of security features WAP2.0 has a full set of security protocols.

此外,蓝牙有许多安全特性,WAP2.0有整套的安全协议。

PART 3 ELECRICAL MACHINES AND DEVICES

Chapter 11 Direct-Current Machines

11.1 Introduction

Electric motors and generators are a group of devices used to convert mechanical energy into electrical energy, or electrical energy into mechanical energy, by electromagnetic means. A machine that converts mechanical energy into electrical energy is called a generator, alternator, or dynamo, and a machine that converts electrical energy into mechanical energy is called a motor.

Two related physical principles underlie the operation of generators and motors. The first is the principle of electromagnetic induction discovered by Michael Faraday in 1831. If a conductor is moved through a magnetic field, or if the strength of a stationary conducting loop is made to vary, a current is set up or induced in the conductor. The converse of this principle is that of electromagnetic reaction, first discovered by Andre' Ampere in 1820. If a current is passed through a conductor located in a magnetic field, the field exerts a mechanical force on it. Both motors and generators consist of two basic units, the field and the armature. The armature is usually a laminated soft-iron core around which conducting wires are wound in coils. Electric machines are classified as DC (Direct Current) and AC (Alternating Current) machines as well as according to their stator and rotor constructions as shown in Fig. 11-1.

Equation $e = d\lambda/dt$, can be used to determine the voltages induced by time-varying magnetic fields. Electromagnetic energy conversion occurs when changes in the flux linkage λ result from mechanical motion. In rotating machines, voltages are generated in windings or groups of coils by rotating these windings mechanically through a magnetic field, by mechanically rotating a magnetic field past the winding, or by designing the magnetic circuit so that the reluctance varies with rotation of the rotor.[1] By any of these methods, the flux linking a specific coil is changed cyclically, and a time-varying voltage is generated.

In most rotating machines, the stator and rotor are made of electrical steel sheet, and the windings are installed in slots on these structures. The use of such high-permeability material maximizes the coupling between the coils and increases the magnetic energy density associated with the

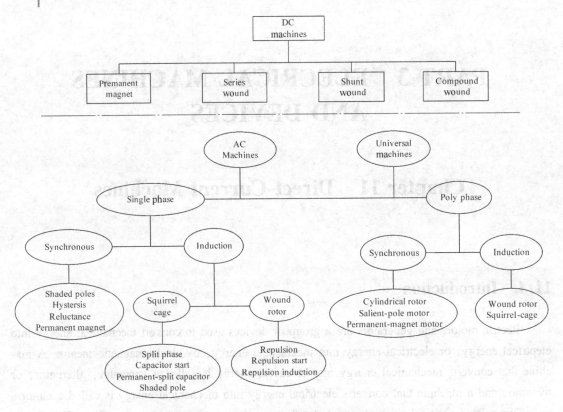

Fig. 11-1 Classification of Electric Machines

electromechanical interaction. It also enables the machine designer to shape and distribute the magnetic fields according to the requirements of each particular machine design. The time-varying flux present in the armature structures of these machines tends to induce currents, known as eddy currents, in the electrical steel. Eddy currents can be a large source of loss in such machines and can significantly reduce machine performance. In order to minimize the effects of eddy currents, the armature structure is typically built from thin laminations of electrical steel which are insulated from each other.

In some machines, such as variable reluctance machines and stepper motors, there are no windings on the rotor. Operation of these machines depends on the nonuniformity of air-gap reluctance associated with variations in rotor position in conjunction with time-varying currents applied to their stator windings.[2] In such machines, both the stator and rotor structures are subjected to time-varying magnetic flux and, as a result, both may require lamination to reduce eddy-current losses.

Rotating electric machines take many forms and are known by many names: DC, synchronous, permanent-magnet, induction, variable reluctance, hysteresis, brushless, and so on. Although these machines appear to be quite dissimilar, the physical principles governing their behavior are quite similar, and it is often helpful to think of them in terms of the same physical picture. For example, analysis of a DC machine shows that associated with both the rotor and the stator are magnetic flux distributions which are fixed in space and that the torque-producing characteristic of the DC machine stems from the tendency of these flux distributions to align.[3] An induction machine, in spite of many fundamental differences, works on exactly the same principle; one can identify flux distributions associated with the rotor and stator. Although they are not stationary but rather rotate in synchronism, just as in a DC motor they are displaced by a constant angular separation, and torque is produced by the tendency of these flux distribution to align.

11.2 Basic Structural Feature

The physical structure of the DC machine consists of two parts: the stator and the rotor, as shown in Fig. 11-2. The stationary part consists of the frame, and the pole pieces, which project inward and provide a path for the magnetic flux. The ends of the pole pieces that are near the rotor spread out over the rotor surface to distribute its flux evenly over the rotor surface. These ends are called the pole shoes. The exposed surface of a pole shoe is called a pole face, and the distance between the pole face and the rotor is the air gap. Two principal windings on a DC machine: ① the armature windings: the windings in which a voltage is induced (rotor); ② the field windings: the windings that produce the main magnetic flux (stator). Note: because the armature winding is located on the rotor, a DC machine's rotor is sometimes called an armature.

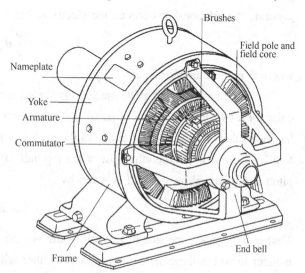

Fig. 11-2 A Simplified Diagram of a DC Machine

11.2.1 Armature Winding

In general, the term armature winding is used to refer to a winding or a set of windings on a rotating machine which carry AC currents. In AC machines such as synchronous or induction machines, the armature winding is typically on the stationary portion of the motor referred to as the stator, in which case these windings may also be referred to as stator windings.[4] In a DC machine, the armature winding is found on the rotating member, referred to as the rotor. As we will see, the armature winding of a DC machine consists of many coils connected together to form a closed loop. A rotating mechanical contact is used to supply current to the armature winding as the rotor rotates.

For small machines, the coils are directly wound into the armature slots using automatic winders. In large machines, the coils are preformed and then inserted into the armature slots. Each coil consists of a number of turns of wire, each turn taped and insulated from the other turns and form the rotor slot. Each side of the turn is called a conductor. The number of conductors on a machine's armature is given by

$$Z = 2CN_c \tag{11-1}$$

where Z is the number of conductors on rotor; C is the number of coils on rotor; N_c is the number of turns per coil.

Since the voltage generated in the conductor under the South Pole opposite the voltage generated in the conductor under the North Pole, the coil span is made equal to 180 electrical degrees, i.e., one pole pitch. In a 2-pole machine 180 electrical degrees is equal to 180 mechanical degrees, whereas in a 4-pole machine 180 electrical degrees is equal to 90 mechanical degrees. In general, the relationship between the electrical angle θ_e and mechanical angle θ_m is given by

$$\theta_e = \frac{P}{2}\theta_m \tag{11-2}$$

where P is the number of poles.

Armature winding of a DC machine is always closed and of double layer type. The number of coils for a two-layer winding is equal to the number of armature slots. Thus, each armature slot has two sides of two different coils. In the preformed coils, one side of a coil is placed at the bottom half of a slot and the other side at the top half. If S is the number of slots in the rotor, the coil pitch is designated by y which is given by

$$y = \text{Integer value of } \left(\frac{S}{P}\right) \tag{11-3}$$

That is if we place one side of the coil in slot m, the other side must be inserted in slot $m + y$. The manner in which the coils are connected together will form the type of armature winding. Let E_c and I_c be the voltage and current per coil. If a is the number of parallel paths between brushes, then the number of series conductor per parallel path between brushes is Z/a. Hence the average generated emf and the armature current are given by

$$E = \frac{Z}{a}E_c$$
$$I_a = aI_c \tag{11-4}$$

1. Simplex Lap Winding

In one type known as the simplex lap winding the end of one coil is connected to the beginning of the next coil with the two ends of each coil coming out at adjacent commutator segments.[5] For a progressive lap winding the commutator pitch $y_c = 1$. A typical coil of N_c turns for a simplex lap winding is shown in Fig. 11-3. In the simplex lap winding the number of parallel path between brushes is equal to the number of poles, i.e., $p = a$. This type of winding is used for low-voltage, high-current applications.

2. Simplex Wave Winding

As we have seen in the lap winding, the two ends of a coil are connected to adjacent commutator segments ($y_c = 1$). In the wave winding, the two ends of a coil are connected to the commutator segments that are approximately 360 electrical degrees apart (i. e., 2-pole pitch). This way all the coils carrying current in the same direction are connected in series. Therefore, there are only two parallel paths between the brushes, i. e., $a = 2$ independent of the number of poles. [6] This type of winding is used for low-current, high-voltage applications.

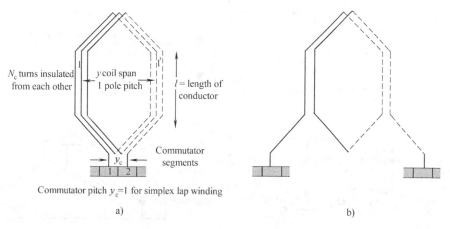

Fig. 11-3
a) A Typical Coil for Lap Winding with $y_c = 1$
b) For Wave Winding

11.2.2 Field Winding

This is an exciting system which may be an electrical winding or a permanent magnet and which is located on the stator and is excited by direct current. The outstanding advantages of DC machines arise from the wide variety of operating characteristics which can be obtained by selection of the method of excitation of the field windings. [7] Various connection diagrams are shown in Fig. 11-4. The method of excitation profoundly influences both the steady-state characteristics and the dynamic behavior of the machine in control systems.

The connection diagram of a separately-excited generator is given in Fig. 11-4a. The required field current is a very small fraction of the rated armature current; on the order of 1 to 3 percent in the average generator. A small amount of power in the field circuit may control a relatively large amount of power in the armature circuit; i. e., the generator is a power amplifier. Separately-excited generators are often used in feedback control systems when control of the armature voltage over a wide range is required. [8] The field windings of self-excited generators may be supplied in three different ways. The field may be connected in series with the armature (Fig. 11-4b), resulting in a series generator. The field may be connected in shunt with the armature (Fig. 11-4c), resulting in a shunt generator, or the field may be in two sections (Fig. 11-4d), one of which is connected in series and the other in shunt with the armature, resulting in a compound generator. [9] The field current of a series generator is the same as the load current, so that the air-gap flux and hence the voltage vary widely with load. As a consequence, series generators are not often used. The voltage of shunt generators drops off somewhat with load, but not in a manner that is objectionable for many purposes. Compound generators are normally connected so that the mmf of the series winding aids that of the shunt winding. The advantage is that through the action of the series winding the flux per pole can increase with load, resulting in a voltage output which is nearly constant or which even rises somewhat as load increases.[10]

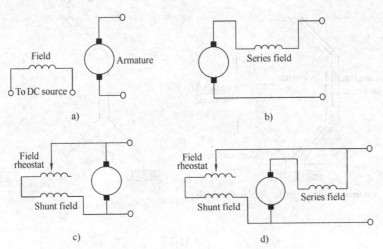

Fig. 11-4 Field-Circuit Connections of DC Machines
a) Separate Excitation   b) Series   c) Shunt   d) Compound

11.2.3 Commutators

A commutator typically consists of a set of copper segments, fixed around part of the circumference of the rotor, and a set of spring-loaded brushes fixed to the stationary frame of the machine.[11] The external source of current (for a motor) or electrical load (for a generator) is connected to the brushes. For small equipment the commutator segments can be stamped from sheet metal. For very large equipment the segments are made from a copper casting that is then machined into the final shape. Fig. 11-5 shows cross-section of a commutator that can be disassembled for repair.

Each conducting segment on the armature of the commutator is insulated from adjacent segments. Initially when the technology was first developed, mica was used as an insulator between commutation segments. Later materials research into polymers brought the development of plastic spacers which are more durable and

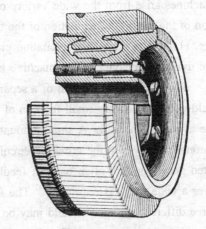

Fig. 11-5 Cross-Section of a Commutator That Can Be Disassembled for Repair

less prone to cracking, and have a higher and more uniform breakdown voltage than mica.[12]

The segments are held onto the shaft using a dovetail shape on the edges or underside of each segment, using insulating wedges around the perimeter of each commutation segment. Due to the high cost of repairs, for small appliance and tool motors the segments are typically crimped permanently in place and cannot be removed; when the motor fails it is simply discarded and replaced. On very large industrial motors it is economical to be able to replace individual damaged segments, and so the end-wedge can be unscrewed and individual segments removed and replaced.

Commutator segments are connected to the coils of the armature, with the number of coils (and commutator segments) depending on the speed and voltage of the machine. Large motors may have hundreds of segments. Friction between the segments and the brushes eventually causes wear to both surfaces. Carbon brushes, being made of a softer material, wear faster and may be designed to be replaced easily without dismantling the machine. Older copper brushes caused more wear to the commutator, causing deep grooving and notching of the surface over time. The commutator on small motors (say, less than a kilowatt rating) is not designed to be repaired through the life of the device. On large industrial equipment, the commutator may be re-surfaced with abrasives, or the rotor may be removed from the frame, mounted in a large metal lathe, and the commutator resurfaced by cutting it down to a smaller diameter.[13] The largest of equipment can include a lathe turning attachment directly over the commutator.

11.2.4 Brushes

Early in the development of dynamos and motors, copper brushes were used to contact the surface of the commutator. However, these hard metal brushes tended to scratch and groove the smooth commutator segments, eventually requiring resurfacing of the commutator. As the copper brushes wear away, the dust and pieces of the brush could wedge between commutator segments, shorting them and reducing the efficiency of the device. Fine copper wire mesh or gauze provided better surface contact with less segment wear, but gauze brushes were more expensive than strip or wire copper brushes. The copper brush was eventually replaced by the carbon brush.

Carbon brushes tend to wear more evenly than copper brushes, and the soft carbon causes far less damage to the commutator segments. There is less sparking with carbon as compared to copper, and as the carbon wears away, the higher resistance of carbon results in fewer problems from the dust collecting on the commutator segments.

Copper and carbon are each better suited for a particular purpose. Copper brushes perform better with very low voltages and high amperage, while carbon brushes are better for high voltage and low amperage. Copper brushes typically carry 150 to 200 A per square inch of contact surface, while carbon only carries 40 A to 70 A per square inch. The higher resistance of carbon also results in a greater voltage drop of 0.8 V to 1.0 V per contact, or 1.6 to 2.0 V across the commutator. Modern rotating machines with commutators now use carbon brushes, which may have copper powder mixed in to improve conductivity. Metallic copper brushes would only be found in toy or very small motors, such as the one illustrated above.

11.3 Effect of Armature MMF

Armature MMF (magnetomotive force) has definite effects on both the space distribution of the air-gap flux and the magnitude of the net flux per pole.[14] The effect on flux distribution is important because the limits of successful commutation are directly influenced; the effect on flux magnitude is important because both the generated voltage and the torque per unit of armature current are influenced thereby. These effects and the problems arising from them are described in this section.

It was shown that the armature MMF wave can be closely approximated by a sawtooth, corresponding to the wave produced by a finely-distributed armature winding or current sheet.[15] For a machine with brushes in the neutral position, the idealized MMF wave is again shown by the dashed sawtooth in Fig. 11-6, in which a positive MMF ordinate denotes flux lines leaving the armature surface. Current directions in all windings other than the main field are indicated by black and cross-hatched bands. Because of the salient-pole field structure found in almost all DC machines, the associated space distribution of flux will not be triangular. The distribution of air-gap flux density with only the armature excited is given by the solid curve of Fig. 11-6. As can readily be seen, it is appreciably decreased by the long air path in the interpolar space.

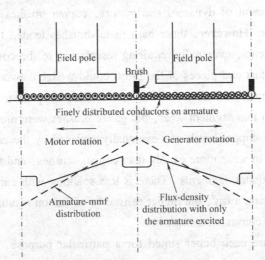

Fig. 11-6 Armature MMF and Flux-Density Distribution with Brushes on Neutral and Only the Armature Excited

The axis of the armature MMF is fixed at 90 electrical degrees from the main-field axis by the brush position. The effect of the armature MMF is seen to be that of creating flux crossing the pole faces; thus its path in the pole shoes crosses the path of the main-field flux.[16] For this reason, armature reaction of this type is called cross-magnetizing armature reaction. It evidently causes a decrease in the resultant air-gap flux density under one half of the pole and an increase under the other half.

When the armature and field windings are both excited, the resultant air-gap flux-density distribution is of the form given by the solid curve of Fig. 11-7. Superimposed on this figure are the flux distributions with only the armature excited (long-dash curve) and only the field excited (short-dash curve). The effect of cross-magnetizing armature reaction in decreasing the flux under one pole tip and increasing it under the other can be seen by comparing the solid and short-dash curves. [17] In general, the solid curve is not the algebraic sum of the two dashed curves because of the nonlinearity of the iron magnetic circuit. Because of saturation of the iron, the flux density is decreased by a greater amount under one pole tip than it is increased under the other. Accordingly, the resultant flux per pole is lower than would be produced by the field winding alone, a consequence known as the demagneti-

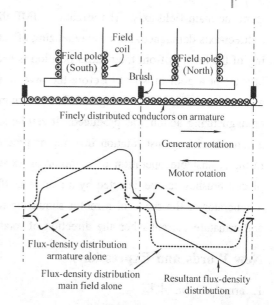

Fig. 11-7 Armature, Main-Field, and Resultant Flux-Density Distributions with Brushed on Neutral

zing effect of cross-magnetizing armature reaction. Since it is caused by saturation, its magnitude is a nonlinear function of both the field current and the armature current. For normal machine operation at the flux densities used commercially, the effect is usually significant, especially at heavy loads, and must often be taken into account in analyses of performance.

The distortion of the flux distribution caused by cross-magnetizing armature reaction may have a detrimental influence on the commutation of the armature current, especially if the distortion becomes excessive. [18] In fact, this distortion is usually an important factor limiting the short-time overload capability of a DC machine. Tendency toward distortion of the flux distribution is most pronounced in a machine, such as a shunt motor, where the field excitation remains substantially constant while the armature MMF may reach very significant proportions at heavy loads. The tendency is least pronounced in a series-excited machine, such as the series motor, for both the field and armature MMF increase with load.

The effect of cross-magnetizing armature reaction can be limited in the design and construction of the machine. The MMF of the main field should exert predominating control on the air-gap flux, so that the condition of weak field MMF and strong armature MMF should be avoided. The reluctance of the cross-flux path (essentially the armature teeth, pole shoes, and the air gap, especially at the pole tips) can be increased by increasing the degree of saturation in the teeth and pole faces, by avoiding too small an air gap, and by using a chamfered or eccentric pole face, which increases the air gap at the pole tips. These expedients affect the path of the main flux as well, but the influence on the cross flux is much greater. The best, but also the most expensive, curative measure is to compensate the armature MMF by means of a winding embedded in the pole faces.

If the brushes are not in the neutral position, the axis of the armature MMF wave is not 90°

from the main-field axis. The armature MMF then produces not only cross magnetization but also a direct-axis demagnetizing or magnetizing effect, depending on the direction of brush shift. Shifting of the brushes from the neutral position is usually inadvertent due to incorrect positioning of the brushes or a poor brush fit. Before the invention of interpoles, however, shifting the brushes was a common method of securing satisfactory commutation, the direction of the shift being such that demagnetizing action was produced. It can be shown that brush shift in the direction of rotation in a generator or against rotation in a motor produces a direct-axis demagnetizing MMF which may result in unstable operation of a motor or excessive drop in voltage of a generator. Incorrectly placed brushes can be detected by a load test. If the brushes are on neutral, the terminal voltage of a generator or the speed of a motor should be the same for identical conditions of field excitation and armature current when the direction of rotation is reversed.

New Words and Expressions

1. armature   n. 电枢
2. magnetic flux linkage   磁链
3. reluctance   n. 磁阻
4. electrical steel sheet   电工钢片
5. permeability   n. 磁导率
6. lamination   n. 叠片
7. magnetic flux   磁通
8. hysteresis   n. 磁滞
9. pole pieces   极靴，极片
10. pole shoes   极靴
11. winder   n. 绕线机
12. conductor   n. 导条
13. magnetomotive force (MMF)   磁动势
14. coil pitch   线圈节距
15. electromotive force (EMF)   电动势
16. simplex lap winding   单叠绕组
17. simplex wave winding   单波绕组
18. permanent magnet   永磁体
19. in series with   与……串联
20. in shunt with   与……并联
21. commutator   n. 换向器
22. commutator segment   换向片
23. polymer   n. 聚合体；多聚物
24. mica   n. 云母片
25. dovetail   n. 楔形；燕尾形
26. perimeter   n. 周边
27. abrasive   n. 研磨剂
28. frame   n. 构架；机座；托架
29. gauze   n. 网纱；纱布
30. sawtooth   n. 锯齿
31. detrimental   adj. 不利的
32. pole tip   极尖

Notes

[1] In rotating machines, voltages are generated in windings or groups of coils by rotating these windings mechanically through a magnetic field, by mechanically rotating a magnetic field past the winding, or by designing the magnetic circuit so that the reluctance varies with rotation of the rotor.

在旋转电动机中，绕组或线圈组中产生的电压可以通过转动处于磁场中的绕组获得，或者通过旋转线圈所处的磁场或通过设计磁路使磁阻随着转子转动而变化来获得。

[2] Operation of these machines depends on the nonuniformity of air-gap reluctance associated with variations in rotor position in conjunction with time-varying currents applied to their sta-

tor windings.

这些电动机的运行依赖于转子位置变化时引起的气隙磁阻的不均匀与定子绕组时变电流的相互作用。

[3] For example, analysis of a DC machine shows that associated with both the rotor and the stator are magnetic flux distributions which are fixed in space and that the torque-producing characteristic of the dc machine stems from the tendency of these flux distributions to align.

例如，对直流电动机的分析表明：与定子和转子都有关联的是磁通分布，其在空间中是固定的以及直流电动机的扭矩特性，其源于用于校准的磁通分布趋势。

[4] In AC machines such as synchronous or induction machines, the armature winding is typically on the stationary portion of the motor referred to as the stator, in which case these windings may also be referred to as stator windings.

交流电动机如同步电动机或感应电动机，电枢绕组典型的放置位置是电动机的静止部分，即定子，因此这些绕组称为定子绕组。

[5] In one type known as the simplex lap winding the end of one coil is connected to the beginning of the next coil with the two ends of each coil coming out at adjacent commutator segments.

一种称作单叠绕组，线圈的末端与下一个绕组的首端相连。每一个线圈的两端分别连接在两个相邻的换向片上。

[6] Therefore, there are only two parallel paths between the brushes, i.e., $a=2$ independent of the number of poles.

因此，对于单波绕组来说，电刷之间只有两个并联支路即 $a=2$，与主极数无关。

[7] The outstanding advantages of DC machines arise from the wide variety of operating characteristics which can be obtained by selection of the method of excitation of the field windings.

直流电动机最显著的优点在于其宽的运行特性，而励磁绕组的励磁方式决定了其具体的运行特性。

[8] Separately-excited generators are often used in feedback control systems when control of the armature voltage over a wide range is required.

当需要在大范围内控制电枢电压时，他励发电机经常在反馈控制系统中采用。

[9] The field may be connected in shunt with the armature (Fig. 11-4c), resulting in a shunt generator, or the field may be in two sections (Fig. 11-4d), one of which is connected in series and the other in shunt with the armature, resulting in a compound generator.

励磁绕组与电枢绕组并联（Fig. 11-4c）称作并励发电机。如果励磁绕组由两部分组成（Fig. 11-4d），一个与电枢绕组串联而另一个与之并联，则称作复励发电机。

[10] The advantage is that through the action of the series winding the flux per pole can increase with load, resulting in a voltage output which is nearly constant or which even rises somewhat as load increases.

复励发电机的优点在于，通过与电枢绕组串联的励磁绕组的作用，每极磁通随负载增加而增加从而导致输出电压接近恒定有时甚至随负载增加而增加。

[11] A commutator typically consists of a set of copper segments, fixed around part of the circumference of the rotor, and a set of spring-loaded brushes fixed to the stationary frame of the machine.

一个换向器由多片铜换向片组成并固定在转子上，电刷再通过压紧弹簧固定在电动机静止框架上。

[12] Later materials research into polymers brought the development of plastic spacers which are more durable and less prone to cracking, and have a higher and more uniform breakdown voltage than mica.

后来，关于聚合物材料的研究研发出了塑料垫片，比起云母它们更加耐用，更不容易破裂，并且具有更加均匀的击穿电压。

[13] On large industrial equipment, the commutator may be re-surfaced with abrasives, or the rotor may be removed from the frame, mounted in a large metal lathe, and the commutator re-surfaced by cutting it down to a smaller diameter.

在大型工业设备上，换向器表面需要用磨料重新磨平，或者将转子从电动机中取出固定在大型机床上，通过切削的方法使换向片直径变小从而达到使表面平滑的目的。

[14] Armature MMF (magnetomotive force) has definite effects on both the space distribution of the air-gap flux and the magnitude of the net flux per pole.

电枢磁动势对气隙磁通的空间分布和每极净磁通大小都有明显的影响。

[15] It was shown that the armature MMF wave can be closely approximated by a sawtooth, corresponding to the wave produced by a finely-distributed armature winding or current sheet.

由此可见，电枢磁动势波可以近似看作一个锯齿波，这个波由一个理想分布的电枢绕组或电流片产生。

[16] The effect of the armature MMF is seen to be that of creating flux crossing the pole faces; thus its path in the pole shoes crosses the path of the main-field flux.

电枢磁动势的影响可以看作是它建立了一个穿过极面的磁通，这个磁通在极靴中的路径与主磁通的路径相交。

[17] The effect of cross-magnetizing armature reaction in decreasing the flux under one pole tip and increasing it under the other can be seen by comparing the solid and short-dash curves.

通过比较图中的实线和短虚线可以看出，交轴电枢反应的作用在于减小一个极尖下磁通而增加另一个极尖磁通。

[18] The distortion of the flux distribution caused by cross-magnetizing armature reaction may have a detrimental influence on the commutation of the armature current, especially if the distortion becomes excessive.

如果由交轴电枢反应引起的磁通分布的畸变程度很大，那么对电枢电流的换向有不利的影响。

Chapter 12 Three-Phase Induction Motors

12.1 Introduction

An AC machine is an electric machine that is driven by an alternating current. It consists of two basic parts, an outside stationary stator having coils supplied with alternating current to produce a rotating magnetic field, and an inside rotor attached to the output shaft that is given a torque by the rotating field.

There are two types of AC machines, depending on the type of rotor used. The first is the synchronous machine, which rotates exactly at the supply frequency or a submultiple of the supply frequency. The magnetic field on the rotor is either generated by current delivered through slip rings or by a permanent magnet. The second type is the induction machine, which turns slightly slower than the supply frequency. The magnetic field on the rotor of this motor is created by an induced current.

In induction machines, alternating currents are applied directly to the stator windings. Rotor currents are then produced by induction. The induction machine may be regarded as a generalized transformer in which electric power is transformed between rotor and stator together with a change of frequency and a flow of mechanical power. [1] Although the induction motor is the most common of all motors, it is seldom used as a generator; its performance characteristics as a generator are unsatisfactory for most applications, although in recent years it has been found to be well suited for wind-power applications. The induction machine may also be used as a frequency changer.

In the induction motor, the stator windings are essentially the same as those of a synchronous machine. However, the rotor windings are electrically short-circuited and frequently have no external connections; currents are induced by transformer action from the stator winding. Several three-phase AC induction motors are shown in Fig. 12-1. The rotor "windings" are actually solid aluminum bars which are cast into the slots in the rotor and which are shorted together by cast aluminum rings at each end of the rotor. [2] This type of rotor construction results in induction motors which are relatively inexpensive and highly reliable, factors contributing to their immense popularity and widespread application.

Fig. 12-1 Three-Phase AC Induction Motors

The armature flux in the induction motor leads that of the rotor and produces an electromechanical torque. In fact, we will see that, just as in a synchronous machine, the rotor and stator fluxes rotate in synchronism with each other and that torque is related to the relative displacement between them. [3] However, unlike a synchronous machine, the rotor of an induction machine does not itself rotate synchronously; it is the "slipping" of the rotor with respect to the synchronous armature flux that gives rise to the induced rotor currents and hence the torque. Induction motors operate at speeds less than the synchronous mechanical speed.

12.2 Construction of Three-Phase Induction Motors

Where a polyphase electrical supply is available, the three-phase AC induction motor is commonly used, especially for higher-powered motors. Like all rotating machines, the AC induction motor consists of two parts—stator and rotor. In the stator, the winding used is a balanced three-phase one, which means that the number of turns in each phase, connected in star/delta, is equal. The windings of the three phases are placed 120° (electrical) apart, the mechanical angle between the adjacent phases being $(2 \times 120°)/p$, where p is the number of poles. For a 4-pole stator, the mechanical angle between the winding of the adjacent phases, is $(2 \times 120°)/4 = 120°/2 = 60°$. The conductors, mostly multi-turn, are placed in the slots, which may be closed, or semi-closed, to keep the leakage inductance low. The start and return parts of the winding are placed nearly 180°, or $(180° - \beta)$ apart. The angle of short chording β is nearly equal to 30°, or close to that value. The short chording results in reducing the amount of copper used for the winding, as the length of the conductor needed for overhang part is reduced.

The section of the stampings used for both stator and rotor is shown in Fig. 12-2. The core is needed below the teeth to reduce the reluctance of the magnetic path, which carries the flux in the motor. The stator is kept normally inside a support.

There are two types of rotors used in induction motors, squirrel cage and wound (slip-ring) one. The cage rotor (Fig. 12-3a) is mainly used, as it is cheap, rugged and needs little or no maintenance. It consists of copper bars placed in the slots of the rotor, short circuited at the two ends by end rings,

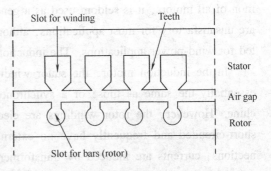

Fig. 12-2 Section for Stamping of Stator and Rotor in Induction Motor

brazed with the bars. The name is derived from the similarity between this rings-and-bars winding and a hamster wheel (presumably similar wheels exist for pet squirrels). [4] The core of the rotor is built of a stack of iron laminations (Fig. 12-3b). Fig. 12-3c shows only three laminations of the stack but many more are used. The currents in the bars of a cage rotor, inserted inside the stator, follow the pattern of currents in the stator winding, when the motor develops torque, such that the number of poles in the rotor is same as that in the stator. If the stator winding of motor is changed, with the number of poles for the new one being different from the earlier one, the cage rotor used need not be changed, thus, can be same, as the current pattern in the rotor bars changes. But the number of poles in the rotor due to the above currents in the bars is same as the number of poles in the new stator winding. The only problem here is that the equivalent resistance of the rotor is constant. So, at the design stage, the value is so chosen, so as to obtain a certain value of the starting torque, and also the slip at full load torque is kept within limits as needed.

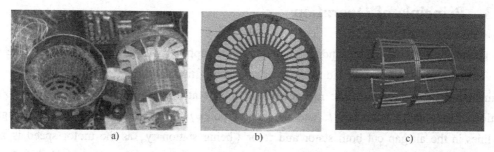

Fig. 12-3 Disassembled Induction Motor with Squirrel Cage Rotor on Its Shaft
a) Squirrel Cage Rotor   b) Rotor Laminations   c) Diagram of the Squirrel-Cage

The other type of rotor, i.e., a wound rotor (slip ring) used has a balanced three-phase winding (Fig. 12-4), being same as the stator winding, but the number of turns used depends on the voltage in the rotor. The three ends of the winding are brought at the three slip-rings, at which points external resistance can be inserted to increase the starting torque requirement. Other three ends are shorted inside. The motor with additional starting resistance is costlier, as this type of rotor is itself costlier than the cage rotor of same power rating, and additional cost of the starting resistance is incurred to increase the starting torque as required. But the slip at full load torque is lower than that of a cage rotor with identical rating, when no additional resistance is used, with direct short-circuiting at the three slip-ring terminals. For a wound (slip-ring) rotor, the rotor winding must be designed for same the number of poles as used for the stator winding. If the number of poles in the rotor winding is different from that of poles in the stator winding, no torque will be developed in the motor. It may be noted that this was not the case with cage rotor, as explained earlier.

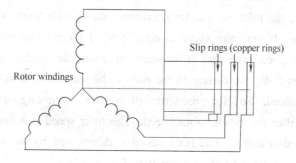

Fig. 12-4 Wound Rotor (Slip Ring) of Induction Motor

The wound rotor in Fig. 12-4 is shown as star-connected, whereas the rotor windings can also be connected in delta, which can be converted into its equivalent star configuration. This shows that the rotor need not always be connected in star as shown. The number of rotor turns changes, as the delta-connected rotor is converted into star-connected equivalent. This point may be kept in mind, while deriving the equivalent circuit, if the additional resistance (being in star) is connected through the slip rings, in series with the rotor winding.

12.3 Principle of Operation

The balanced three-phase winding of the stator is supplied with a balanced three-phase voltage. The current in the stator winding produces a rotating magnetic field, the magnitude of which remains constant. The axis of the magnetic field rotates at a synchronous speed $n_s = (2f)/p$, a function of the supply frequency f, and number of poles p in the stator winding. The magnetic flux lines in the air gap cut both stator and rotor (being stationary, as the motor speed is zero) conductors at the same speed. The EMFs in both stator and rotor conductors are induced at the same frequency, i.e. supply frequency, with the number of poles for both stator and rotor windings (assuming wound one) being same.[5] The stator conductors are always stationary, with the frequency in the stator winding being same as supply frequency. As the rotor winding is short-circuited at the slip-rings, current flows in the rotor windings. The electromagnetic torque in the motor is in the same direction as that of the rotating magnetic field, due to the interaction between the rotating flux produced in the air gap by the current in the stator winding, and the current in the rotor winding.[6] This is as Lenz's law, as the developed torque is in such direction that it will oppose the cause, which results in the current flowing in the rotor winding. This is irrespective of the rotor type used cage or wound one, with the cage rotor, with the bars short-circuited by two end-rings, is considered equivalent to a wound one.[7] The current in the rotor bars interacts with the air-gap flux to develop the torque, irrespective of the number of poles for which the winding in the stator is designed. Thus, the cage rotor may be termed as universal one. The induced EMF and the current in the rotor are due to the relative velocity between the rotor conductors and the rotating flux in the air-gap, which is maximum, when the rotor is stationary ($n_r = 0$).[8] As the rotor starts rotating in the same direction, as that of the rotating magnetic field due to production of the torque as stated earlier, the relative velocity decreases, along with lower values of induced EMF and current in the rotor. If the rotor speed is equal that of the rotating magnetic field, which is termed as synchronous speed, and also in the same direction, the relative velocity is zero, which causes both the induced EMF and current in the rotor to be reduced to zero. Under this condition, torque will not be produced. So, for production of positive (motoring) torque, the rotor speed must always be lower than the synchronous speed. The rotor speed is never equal to the synchronous speed in an induction motor. The rotor speed is determined by the mechanical load on the shaft and the total rotor losses, mainly comprising of copper loss.

The difference between the synchronous speed and rotor speed, expressed as a ratio of the synchronous speed, is termed as 'slip' in an induction motor. So, slip s is

$$s = \frac{n_s - n_r}{n_s} = 1 - \frac{n_r}{n_s} \text{ or, } n_r = (1-s)n_s \qquad (12\text{-}1)$$

where, n_s and n_r are synchronous and rotor speeds in r/s. Normally, for torques varying from no-load ($\approx$ zero) to full load value, the slip is proportional to torque. The slip at full load is 4%-5%.

An alternative explanation for the production of torque in a three-phase induction motor is given here, using two rules (right hand and left hand) of Fleming. The stator and rotor, along with air-gap, is shown in Fig. 12-5a. Both stator and rotor is shown there as surfaces, but without the slots. Also shown is the path of the flux in the air gap. This is for a section, which is under North Pole, as the flux lines move from stator to rotor. The rotor conductor shown in the figure is at rest, i. e., zero speed (stand-still). The rotating magnetic field moves past the conductor at synchronous speed in the clockwise direction. Thus, there is relative movement between the flux and the rotor conductor. Now, if the magnetic field, which is rotating, is assumed to be at stand-still as shown in Fig. 12-5b, the conductor will move in the direction shown. So, an EMF is induced in the rotor conductor as Faraday's law, due to change in flux linkage. The direction of the induced EMF as shown in the figure can be determined using Fleming's right hand rule.

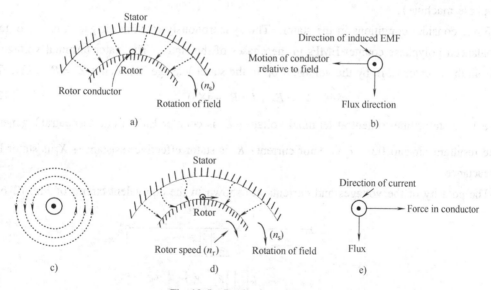

Fig. 12-5 Production of Torque

As described earlier, the rotor bars in the cage rotor are short circuited via end rings. Similarly, in the wound rotor, the rotor windings are normally short-circuited externally via the slip rings.[9] In both cases, as EMF is induced in the rotor conductor (bar), current flows there, as it is short circuited. The flux in the air gap, due to the current in the rotor conductor is shown in Fig. 12-5c. The flux pattern in the air gap, due to the magnetic fields produced by the stator windings and the current carrying rotor conductor, is shown in Fig. 12-5d. The flux lines bend as shown there. The property of the flux lines is to travel via shortest path as shown in Fig. 12-5a. If the flux lines try to move to form straight line, then the rotor conductor has to move in the direction of the rotating magnetic field, but not at the same speed, as explained earlier.[10] The current carrying rotor conductor and the direction of flux are shown in Fig. 12-5e. It is known that force is produced on the conductor carrying current, when it is placed in a magnetic field. The direction of the force on the rotor conductor is obtained by using Fleming's left hand rule, being same as that of the rotating magnetic field. Thus, the rotor experiences a motoring torque in the same direction as that of the rotating magnetic field. This briefly describes how torque is produced in a three-phase induction motor.

12.4 Equivalent Circuit

In this derivation, only machines with symmetric polyphase windings excited by balanced polyphase voltages are considered. As in many other discussions of polyphase devices, it is helpful to think of three-phase machines as being Y-connected, so that currents are always line values and voltages always line-to-neutral values. In this case, we can derive the equivalent circuit for one phase, with the understanding that the voltages and currents in the remaining phases can be found simply by an appropriate phase shift of those of the phase under study (±120° in the case of a three-phase machine).

First, consider conditions in the stator. The synchronously-rotating air-gap flux wave generates balanced polyphase counter EMFs in the phases of the stator. The stator terminal voltage differs from the counter EMF by the voltage drop in the stator leakage impedance $Z_1 = R_1 + jX_1$. Thus

$$\hat{V}_1 = \hat{E}_2 + \hat{I}_1(R_1 + jX_1) \tag{12-2}$$

Where $\hat{V}_1$ is stator line-to-neutral terminal voltage; $\hat{E}_2$ is counter EMF (line-to-neutral) generated by the resultant air-gap flux, $\hat{I}_1$ is stator current; R_1 is stator effective resistance; X_1 is stator leakage reactance.

The polarity of the voltages and currents are shown in the equivalent circuit of Fig. 12-6.

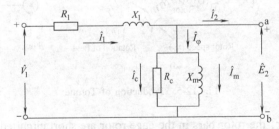

Fig. 12-6 Stator Equivalent Circuit for a Polyphase Induction Motor

The resultant air-gap flux is created by the combined MMF's of the stator and rotor currents. The stator current can be resolved into two components: a load component and an exciting (magnetizing) component. The load component $\hat{I}_2$ produces an MMF that corresponds to the MMF of the rotor current. The exciting component $\hat{I}_\varphi$ is the additional stator current required to create the resultant air-gap flux and is a function of the EMF $\hat{E}_2$. The exciting current can be resolved into a core-loss component $\hat{I}_c$ in phase with $\hat{E}_2$ and a magnetizing component $\hat{I}_m$ lagging $\hat{E}_2$ by 90°. In the equivalent circuit, the exciting current can be accounted for by means of a shunt branch, formed by a core-loss resistance R_c and a magnetizing reactance X_m in parallel, connected across $\hat{E}_2$, as shown in Fig. 12-6. [11] Both R_c and X_m are usually determined at rated stator frequency and for a value of E_2 close to the expected operating value; they are then assumed to remain constant for the small departures of E_2 associated with normal operation of the motor.

The equivalent circuit representing stator phenomena is exactly like that used to represent the primary of a transformer. To complete our model, the effects of the rotor must be incorporated. From the point of view of the stator equivalent circuit of Fig. 12-6, the rotor can be represented by an equivalent impedance Z_2

$$Z_2 = \frac{\hat{E}_2}{\hat{I}_2} \tag{12-3}$$

corresponding to the leakage impedance of an equivalent stationary secondary. To complete the equivalent circuit, we must determine Z_2 by representing the stator and rotor voltages and currents in terms of rotor quantities as referred to the stator. [12]

From the point of view of the primary, the secondary winding of a transformer can be replaced by an equivalent secondary winding having the same number of turns as the primary winding. In a transformer where the turns ratio and the secondary parameters are known, this can be done by referring the secondary impedance to the primary by multiplying it by the square of the primary-to-secondary turns ratio. [13] The resultant equivalent circuit is perfectly general from the point of view of primary quantities.

Similarly, in the case of a polyphase induction motor, if the rotor were to be replaced by an equivalent rotor with a polyphase winding with the same number of phases and turns as the stator but producing the same MMF and air gap flux as the actual rotor, the performance as seen from the stator terminals would be unchanged. This concept, which we will adopt here, is especially useful in modeling squirrel-cage rotors for which the identity of the rotor "phase windings" is in no way obvious.

The rotor of an induction machine is short-circuited, and hence the impedance seen by induced voltage is simply the rotor short-circuit impedance. Consequently the relation between the slip-frequency leakage impedance Z_{2s} of the equivalent rotor and the slip-frequency leakage impedance Z_{rotor} of the actual rotor must be

$$Z_{2s} = \frac{\hat{E}_{2s}}{\hat{I}_{2s}} = N_{eff}^2 \left(\frac{\hat{E}_{rotor}}{\hat{I}_{rotor}} \right) = N_{eff}^2 Z_{rotor} \tag{12-4}$$

where N_{eff} is the effective turns ratio between the stator winding and that of the actual rotor winding. Here the subscript 2s refers to quantities associated with the referred rotor. Thus $\hat{E}_{2s}$ is the voltage induced in the equivalent rotor by the resultant air-gap flux, and $\hat{I}_{2s}$ is the corresponding induced current.

When one is concerned with the actual rotor currents and voltages, the turns ratio N_{eff} must be known in order to convert back from equivalent-rotor quantities to those of the actual rotor. However, for the purposes of studying induction-motor performance as seen from the stator terminals, there is no need for this conversion and a representation in terms of equivalent-rotor quantities is fully adequate. Thus an equivalent circuit based upon equivalent-rotor quantities can be used to represent both coil-wound and squirrel-cage rotors.

Having taken care of the effects of the stator-to-rotor turns ratio, we next must take into account the relative motion between the stator and the rotor with the objective of replacing the actual rotor and its slip-frequency voltages and currents with an equivalent stationary rotor with stator-frequency voltages and currents. Consider first the slip-frequency leakage impedance of the referred rotor.

$$Z_{2s} = \frac{\hat{E}_{2s}}{\hat{I}_{2s}} = R_2 + jsX_2 \tag{12-5}$$

Where R_2 is referred rotor resistance, sX_2 is referred rotor leakage reactance at slip frequency Note that here X_2 has been defined as the referred rotor leakage reactance at stator frequency f_e. Since the actual rotor frequency $f_r = sf_e$, it has been converted to the slip-frequency reactance simply by multiplying by the slip s. The slip-frequency equivalent circuit of one phase of the referred rotor is shown in Fig. 12-7. This is the equivalent circuit of the rotor as seen in the slip-frequency rotor reference frame.

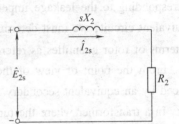

Fig. 12-7 Rotor Equivalent Circuit for a Polyphase Induction Motor at Slip Frequency

We next observe that the resultant air-gap mmf wave is produced by the combined effects of the stator current $\hat{I}_1$ and the equivalent load current $\hat{I}_2$. Similarly, it can be expressed in terms of the stator current and the equivalent rotor current $\hat{I}_{2s}$. These two currents are equal in magnitude since $\hat{I}_{2s}$ is defined as the current in an equivalent rotor with the same number of turns per phase as the stator. Because the resultant air-gap mmf wave is determined by the phasor sum of the stator current and the rotor current of either the actual or equivalent rotor, $\hat{I}_2$ and $\hat{I}_{2s}$ must also be equal in phase (at their respective electrical frequencies) and hence we can write

$$\hat{I}_{2s} = \hat{I}_2 \tag{12-6}$$

Finally, consider that the resultant flux wave induces both the slip-frequency emf induced in the referred rotor $\hat{E}_{2s}$ and the stator counter EMF $\hat{E}_2$. If it were not for the effect of speed, these voltages would be equal in magnitude since the referred rotor winding has the same number of turns per phase as the stator winding. However, because the relative speed of the flux wave with respect to the rotor is s times its speed with respect to the stator, the relation between these emfs is

$$E_{2s} = sE_2 \tag{12-7}$$

We can furthermore argue that since the phase angle between each of these voltages and the resultant flux wave is 90°, then these two voltages must also be equal in a phasor sense at their respective electrical frequencies. Hence

$$\hat{E}_{2s} = s\hat{E}_2 \tag{12-8}$$

Division of Eq. 12-7 by Eq. 12-6 and use of Eq. 12-5 then gives

$$\frac{\hat{E}_{2s}}{\hat{I}_{2s}} = \frac{s\hat{E}_2}{\hat{I}_2} = Z_{2s} = R_2 + jsX_2 \tag{12-9}$$

Division by the slip s then gives

$$Z_2 = \frac{\hat{E}_2}{\hat{I}_2} = \frac{R_2}{s} + jX_2 \tag{12-10}$$

We have achieved our objective. Z_2 is the impedance of the equivalent stationary rotor which appears across the load terminals of the stator equivalent circuit of Fig. 12-6. The final result is shown in the single-phase equivalent circuit of Fig. 12-8. The combined effect of shaft load and rotor resistance appears as a reflected resistance R_2/s, a function of slip and therefore of the mechanical load.[14] The current in the reflected rotor impedance equals the load component $\hat{I}_2$ of stator current; the voltage across this impedance equals the stator voltage $\hat{E}_2$. Note that when rotor currents and voltages are reflected into the stator, their frequency is also changed to stator frequency. All rotor electrical phenomena, when viewed from the stator, become stator-frequency phenomena, because the stator winding simply sees MMF and flux waves traveling at synchronous speed.[15]

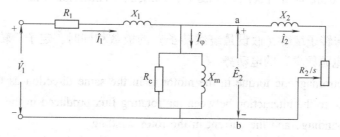

Fig. 12-8 Single-Phase Equivalent Circuit for a Polyphase Induction Motor

New Words and Expressions

1. slip ring 滑环
2. in synchronism with 与……同步
3. hamster *n.* 仓鼠
4. wound rotor 绕线转子
5. squirrel rotor 笼型转子
6. irrespective of 不顾，不考虑
7. no load 空载
8. full load 满载
9. standstill *adj.* 静止的
10. leakage impedance 漏阻抗
11. in phase 同相
12. out of phase 异相
13. polyphase *n.* 多相
14. resultant *adj.* 合成的

Notes

[1] The induction machine may be regarded as a generalized transformer in which electric power is transformed between rotor and stator together with a change of frequency and a flow of mechanical power.

感应电动机可以看作是一个一般化的变压器。在这台"变压器"中，电功率在定子、转子之间传递，并伴随着频率的变化以及机械功率的输出。

[2] The rotor "windings" are actually solid aluminum bars which are cast into the slots in the rotor and which are shorted together by cast aluminum rings at each end of the rotor.

转子绕组实际上是浇铸在转子槽中的实心铝条，这些铝条通过转子端部的铸铝端环短路。

[3] In fact, we will see that, just as in a synchronous machine, the rotor and stator fluxes rotate in synchronism with each other and that torque is related to the relative displacement between them.

实际上，我们将会看到，与同步电动机类似，转子与定子磁通以相同的速度旋转，转矩大小与两者之间的相对位移有关。

[4] The name is derived from the similarity between this rings-and-bars winding and a hamster wheel (presumably similar wheels exist for pet squirrels).

由于这个由导条和端环组成的绕组与仓鼠跑轮（大概与宠物松鼠跑轮类似）的相似性，故命名为笼型转子。

[5] The EMFs in both stator and rotor conductors are induced at the same frequency, i.e. supply frequency, with the number of poles for both stator and rotor windings (assuming wound one) being same.

当定子绕组与转子绕组（假设是绕线转子）的极数相同时，定子、转子导体中会感应出相同频率（即电源频率）的电动势。

[6] The electromagnetic torque in the motor is in the same direction as that of the rotating magnetic field, due to the interaction between the rotating flux produced in the air gap by the current in the stator winding, and the current in the rotor winding.

电动机中的电磁转矩方向与旋转磁场方向同向。此转矩由定子绕组电流在气隙中产生的旋转磁场与转子绕组电流相互作用产生。

[7] This is irrespective of the rotor type used cage or wound one, with the cage rotor, with the bars short-circuited by two end-rings, is considered equivalent to a wound one.

这与所使用的转子类型无关。在笼形转子中，导条通过两个端环短路，这样就与绕线转子等效了。

[8] The induced EMF and the current in the rotor are due to the relative velocity between the rotor conductors and the rotating flux in the air-gap, which is maximum, when the rotor is stationary ($n_r = 0$).

转子中的感应电动势由转子导条与气隙中的旋转磁通相对运动产生。当转子静止时（$n_r = 0$）感应电动势达到最大值。

[9] As described earlier, the rotor bars in the cage rotor are short circuited via end rings. Similarly, in the wound rotor, the rotor windings are normally short-circuited externally via the slip rings.

如前文所述,笼型转子导条经端环短路。类似的,绕线转子的绕组通常通过外部的集电环短路。

[10] If the flux lines try to move to form straight line, then the rotor conductor has to move in the direction of the rotating magnetic field, but not at the same speed, as explained earlier.

如果磁力线试图形成直线,那么转子导体必须转到与旋转磁场相同的方向上,但是两者的速度不会一样,正如前文所解释的那样。

[11] In the equivalent circuit, the exciting current can be accounted for by means of a shunt branch, formed by a core-loss resistance R_c and a magnetizing reactance X_m in parallel, connected across $\hat{E}_2$, as shown in Fig. 12-6.

在等效电路中,通过一个支路来计及励磁电流的影响,这个支路由一个铁耗电阻 R_c 和一个励磁电抗 X_m 并联组成连接在 $\hat{E}_2$ 两端,如图12-6所示。

[12] To complete the equivalent circuit, we must determine Z_2 by representing the stator and rotor voltages and currents in terms of rotor quantities as referred to the stator.

为了完成等效电路,我们必须通过将转子各物理量向定子折算得到折算后的定转子侧的电压电流从而确定阻抗 Z_2。

[13] In a transformer where the turns ratio and the secondary parameters are known, this can be done by referring the secondary impedance to the primary by multiplying it by the square of the primary-to-secondary turns ratio.

在变压器中,匝比和二次侧参数是已知的。而二次侧参数向一次侧折算则可以通过在二次侧阻抗上乘以匝比的平方来实现。

[14] The combined effect of shaft load and rotor resistance appears as a reflected resistance R_2/s, a function of slip and therefore of the mechanical load.

轴端机械负载与转子电阻的影响通过 R_2/s 反映到等效电路中,它是滑差的函数,因此也是机械负载的函数。

[15] All rotor electrical phenomena, when viewed from the stator, become stator-frequency phenomena, because the stator winding simply sees MMF and flux waves traveling at synchronous speed.

当从定子角度来看时,所有转子的电现象就变为定子频率现象,因为定子绕组把磁动势和磁通波形简单地看作以同步的速度传播。

Chapter 13 Synchronous Machines

13.1 Introduction

In synchronous machines, rotor-winding currents are supplied directly from the stationary frame through a rotating contact. A direct current is applied to the rotor winding, which then produces a rotor magnetic field. The rotor is then turned by a prime mover (eg. steam, water, etc.) producing a rotating magnetic field. This rotating magnetic field induces a 3-phase set of voltages within the stator windings of the synchronous generator.

"Field windings" applies to the windings that produce the main magnetic field in a machine, and "armature windings" applies to the windings where the main voltage is induced. For synchronous machines, the field windings are on the rotor, so the terms "rotor windings" and "field windings" are used interchangeably.[1] The rotor of a synchronous generator is a large electromagnet and the magnetic poles on the rotor can either be salient or non salient construction. Non-salient pole rotors are normally used for rotors with 2 or 4 poles rotor, while salient pole rotors are used for 4 or more poles rotor.[2]

Adirect current must be supplied to the field circuit on the rotor. Since the rotor is rotating, a special arrangement is required to get the DC power to its field windings. The common ways are: ①Supply the DC power from an external DC source to the rotor by means of slip rings and brushes; ②Supply the DC power from a special DC power source mounted directly on the shaft of the synchronous generator. Slip rings are metal rings completely encircling the shaft of a machine but insulated from it. One end of the DC rotor winding is tied to each of the 2 slip rings on the shaft of the synchronous machine, and a stationary brush rides on each slip ring. A "brush" is a block of graphite-like carbon compound that conducts electricity freely but has very low friction, hence it doesn't wear down the slip ring. If the positive end of a DC voltage source is connected to one brush and the negative end is connected to the other, then the same DC voltage will be applied to the field winding at all times regardless of the angular position or speed of the rotor.[3] Some problems with slip rings and brushes:

1) They increase the amount of maintenance required on the machine, since the brushes must be checked for wear regularly.

2) Brush voltage drop can be the cause of significant power losses on machines with larger field currents.

Small synchronous machines use slip rings and brushes, and larger machines use brushless exciters are used to supply the DC field current. A brushless exciter is a small AC generator with its field circuit mounted on the stator and its armature circuit mounted on the rotor shaft.[4] The

3-phase output of the exciter generator is rectified to direct current by a 3-phase rectifier circuit also mounted on the shaft of the generator, and is then fed to the main DC field circuit. By controlling the small DC field current of the exciter generator (located on the stator), we can adjust the field current on the main machine without slip rings and brushes. Since no mechanical contacts occur between the rotor and stator, a brushless exciter requires less maintenance. To make the excitation of a generator completely independent of any external power sources, a small pilot exciter can be used. A pilot exciter is a small AC generator with permanent magnets mounted on the rotor shaft and a 3-phase winding on the stator. It produces the power for the field circuit of the exciter, which in turn controls the field circuit of the main machine. If a pilot exciter is included on the generator shaft, then no external electric power is required.

13.2 Principle of Operation

A preliminary picture of synchronous machine performance can be gained by discussing the voltage induced in the armature of the very much simplified *salient-pole* AC synchronous generator shown schematically in Fig. 13-1.[5] The field-winding of this machine produces a single pair of magnetic poles (similar to that of a bar magnet), and hence this machine is referred to as a *two-pole* machine.

With rare exceptions, the armature winding of a synchronous machine is on the stator, and the field winding is on the rotor, as is true for the simplified machine of Fig. 13-1. The field winding is excited by direct current conducted to it by means of stationary carbon brushes which contact rotating *slip rings* or *collector rings*.[6] Practical factors usually dictate this orientation of the two windings: It is advantageous to have the single, low-power field winding on the rotor while having the high-power, typically multiple-phase, and armature winding on the stator.

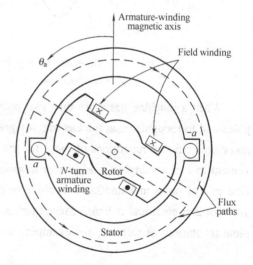

Fig. 13-1 Schematic View of a Simple Two-Pole, Single-Phase Synchronous Generator

The armature winding, consisting here of only a single coil of N turns, is indicated in cross section by the two coil sides a and $-a$ placed in diametrically opposite narrow slots on the inner periphery of the stator of Fig. 13-1. The conductors forming these coil sides are parallel to the shaft of the machine and are connected in series by end connections (not shown in the figure). The rotor is turned at a constant speed by a source of mechanical power connected to its shaft. The armature winding is assumed to be open-circuited and hence the flux in this machine is produced by the field winding alone. Flux paths are shown schematically by dashed lines in Fig. 13-1.

A highly idealized analysis of this machine would assume a sinusoidal distribution of magnetic flux in the air gap. The resultant radial distribution of air-gap flux density B is shown in Fig. 13-2a as a function of the spatial angle θ_a (measured with respect to the magnetic axis of the armature winding) around the rotor periphery. [7] In practice, the air-gap flux-density of practical salient-pole machines can be made to approximate a sinusoidal distribution by properly shaping the pole faces.

As the rotor rotates, the flux-linkages of the armature winding change with time. Under the assumption of a sinusoidal flux distribution and constant rotor speed, the resulting coil voltage will be sinusoidal in time as shown in Fig. 13-2b. The coil voltage passes through a complete cycle for each revolution of the two-pole machine of Fig. 13-1. Its frequency in cycles per second (Hz) is the same as the speed of the rotor in revolutions per second: the electric frequency of the generated voltage is synchronized with the mechanical speed, and this is the reason for the designation "synchronous" machine. [8] Thus a two-pole synchronous machine must revolve at 3,600 revolutions per minute to produce a 60 Hz voltage.

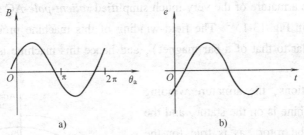

Fig. 13-2 Generated Voltage for the Single-Phase Generator of Fig. 13-1
a) Space Distribution of Flux Density b) Corresponding Waveform

When a machine has more than two poles, it is convenient to concentrate on a single pair of poles and to recognize that the electric, magnetic, and mechanical conditions associated with every other pole pair are repetitions of those for the pair under consideration. [9] For this reason it is convenient to express angles in *electrical degrees* or *electrical radians* rather than in physical units. One pair of poles in a multipole machine or one cycle of flux distribution equals 360 electrical degrees or 2π electrical radians. Since there are $p/2$ complete wavelengths, or cycles, in one complete revolution, it follows, for example, that

$$\theta_{ae} = \left(\frac{p}{2}\right)\theta_a \qquad (13-1)$$

where θ_{ae} is the angle in electrical units and θ_a is the spatial angle. This same relationship applies to all angular measurements in a multipole machine; their values in electrical units will be equal to (poles/2) times their actual spatial values.

The coil voltage of a multipole machine passes through a complete cycle every time a pair of poles sweeps by, or $(p/2)$ times each revolution. The electrical frequency f_e of the voltage generated in a synchronous machine is therefore

$$f_e = \left(\frac{p}{2}\right)\frac{n}{60} \text{ Hz} \qquad (13-2)$$

where n is the mechanical speed in revolutions per minute, and hence $n/60$ is the speed in revolutions per second. The electrical frequency of the generated voltage in radians per second is $\omega_e = (p/2)\omega_m$ where ω_m is the mechanical speed in radians per second.

The rotors shown in Fig. 13-1 have *salient*, or *projecting*, poles with *concentrated windings*. Fig. 13-3 shows diagrammatically a *nonsalient-pole*, or *cylindrical* rotor. The field winding is a two-pole *distributed winding*; the coil sides are distributed in multiple slots around the rotor periphery and arranged to produce an approximately sinusoidal distribution of radial air-gap flux.[10]

The relationship between electrical frequency and rotor speed of Equation 13-2 can serve as a basis for understanding why some synchronous generators have salient-pole rotor structures while others have cylindrical rotors. Most power systems in the world operate at frequencies of either 50 Hz or 60 Hz. A salient-pole construction is char-

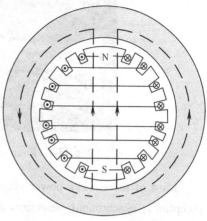

Fig. 13-3 Elementary Two-Pole Cylindrical-Rotor Field Winding

acteristic of hydroelectric generators because hydraulic turbines operate at relatively low speeds, and hence a relatively large number of poles is required to produce the desired frequency; the salient-pole construction is better adapted mechanically to this situation.[11] The rotor of a large hydroelectric generator is shown in Fig. 13-4. Steam turbines and gas turbines, however, operate best at relatively high speeds, and turbine-driven alternators or turbine generators are commonly two- or four-pole cylindrical-rotor machines. The rotors are made from a single steel forging or from several forgings, as shown in Fig. 13-5.

Fig. 13-4 Water-Cooled Rotor of 190-MVA Hydroelectric Generator

Fig. 13-5   Rotor of a Two-Pole 3,600 r/min Turbine Generator

Most of the world's power systems are three-phase systems and, as a result, with very few exceptions, synchronous generators are three-phase machines. For the production of a set of three voltages phase-displaced by 120 electrical degrees in time, a minimum of three coils phase-displaced 120 electrical degrees in space must be used.[12] A simplified schematic view of a three-phase, two-pole machine with one coil per phase is shown in Fig. 13-6a. The three phases are designated by the letters a, b, and c. In an elementary four-pole machine, a minimum of two such sets of coils must be used, as illustrated in Fig. 13-6b; in an elementary multipole machine, the minimum number of coils sets is given by one half the number of poles.

The two coils in each phase of Fig. 13-6b are connected in series so that their voltages add, and the three phases may then be either Y-or Δ-connected. Fig. 13-6c shows how the coils may be interconnected to form a Y connection. Note however, since the voltages in the coils of each phase are identical, a parallel connection is also possible, e.g., coil $(a, -a)$ in parallel with coil $(a', -a')$, and so on.

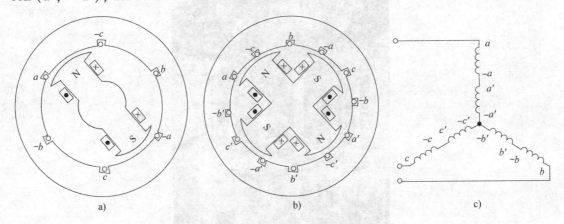

Fig. 13-6   Schematic Views of Three-Phase Generators
a) Two-Pole   b) Four-Pole   c) Y Connection of the Windings

When a synchronous generator supplies electric power to a load, the armature current creates a magnetic flux wave in the air gap which rotates at synchronous speed. This flux reacts with the flux created by the field current, and electromechanical torque results from the tendency of these two magnetic fields to align. [13] In a generator this torque opposes rotation, and mechanical torque must be applied from the prime mover to sustain rotation. This electromechanical torque is the mechanism through which the synchronous generator converts mechanical to electric energy.

The counterpart of the synchronous generator is the synchronous motor. Alternating current is supplied to the armature winding on the stator, and DC excitation is supplied to the field winding on the rotor. The magnetic field produced by the armature currents rotates at synchronous speed. To produce a steady electromechanical torque, the magnetic fields of the stator and rotor must be constant in amplitude and stationary with respect to each other. In a synchronous motor, the steady-state speed is determined by the number of poles and the frequency of the armature current. [14] Thus a synchronous motor operated from a constant-frequency AC source will operate at a constant steady-state speed.

In a motor the electromechanical torque is in the direction of rotation and balances the opposing torque required to drive the mechanical load. The flux produced by currents in the armature of a synchronous motor rotates ahead of that produced by the field, thus pulling on the field (and hence on the rotor) and doing work. [15] This is the opposite of the situation in a synchronous generator, where the field does work as its flux pulls on that of the armature, which is lagging behind. In both generators and motors, an electromechanical torque and a rotational voltage are produced. These are the essential phenomena for electromechanical energy conversion.

New Words and Expressions

1. prime mover   原动机
2. salient rotor motor   凸极电动机
3. cylindrical rotor motor   隐极电动机
4. graphite   *n.* 石墨
5. regardless of   不顾；不惜
6. exciter   *n.* 励磁机
7. rectifier circuit   整流器电路
8. pilot exciter   副励磁机
9. electrical degree   电角度
10. spatial angle   空间角
11. air-gap flux   气隙磁通
12. hydroelectric generator   水力发电机
13. steam turbine   汽轮机
14. steel forging   钢锻品
15. in parallel with   与……平行

Notes

[1] For synchronous machines, the field windings are on the rotor, so the terms "rotor windings" and "field windings" are used interchangeably.

对于同步电机来说，由于励磁绕组在转子上，因此"转子绕组"和"励磁绕组"这两个术语在使用时可以互换。

[2] Non-salient pole rotors are normally used for rotors with 2 or 4 poles rotor, while salient pole rotors are used for 4 or more poles rotor.

隐极转子通常采用2或4极，而凸极转子通常是4极或更多极。

[3] If the positive end of a DC voltage source is connected to one brush and the negative end is connected to the other, then the same DC voltage will be applied to the field winding at all times regardless of the angular position or speed of the rotor.

如果直流电源的正、负端分别连接在两个电刷上，那么不管转子的角位置或是转速如何，励磁绕组两端的电压总是与直流电源电压相同。

[4] A brushless exciter is a small AC generator with its field circuit mounted on the stator and its armature circuit mounted on the rotor shaft.

一台无刷励磁机是一种励磁电路安装在定子上而电枢电路在转子转轴上的小型交流发电机。

[5] A preliminary picture of synchronous machine performance can be gained by discussing the voltage induced in the armature of the very much simplified *salient-pole* AC synchronous generator shown schematically in Fig. 13-1.

如图 13-1 所示，通过讨论一个非常简化的凸极交流同步发电机电枢绕组上的感应电压可以初步了解同步发电机的性能。

[6] The field winding is excited by direct current conducted to it by means of stationary carbon brushes which contact rotating *slip rings or collector rings*.

通过与旋转的滑环或集电环接触的电刷获取直流电流来激励励磁绕组。

[7] The resultant radial distribution of air-gap flux density B is shown in Fig. 13-2a as a function of the spatial angle θ_a (measured with respect to the magnetic axis of the armature winding) around the rotor periphery.

图 13-2a 显示了合成的气隙磁密 B 的径向分布与空间角 θ_a 的函数关系。

[8] Its frequency in cycles per second (Hz) is the same as the speed of the rotor in revolutions per second: the electric frequency of the generated voltage is synchronized with the mechanical speed, and this is the reason for the designation "synchronous" machine.

它的频率 (Hz) 与转子每秒的转数相同：所发出电压的电气频率与机械转速同步，这就是为什么我们称它为"同步"发电机的原因。

[9] When a machine has more than two poles, it is convenient to concentrate on a single pair of poles and to recognize that the electric, magnetic, and mechanical conditions associated with every other pole pair are repetitions of those for the pair under consideration.

当电机的极数超过两极，为了方便可以仅分析一对极下的电、磁以及机械关系，因为其他极对下的情况仅仅是这一对极的重复。

[10] The field winding is a two-pole distributed winding; the coil sides are distributed in multiple slots around the rotor periphery and arranged to produce an approximately sinusoidal distribution of radial air-gap flux.

励磁绕组是一个两极分布绕组；线圈边分布在转子周边的多级槽中，产生一个近似正弦分布的径向气隙磁通。

[11] A salient-pole construction is characteristic of hydroelectric generators because hydraulic turbines operate at relatively low speeds, and hence a relatively large number of poles is required to produce the desired frequency; the salient-pole construction is better adapted mechanical-

ly to this situation.

水轮发电机采用凸极结构的转子因为水轮机的运行速度较低，因此为了产生一定频率的交流电力则需要较多的极数；凸极的结构在机械上更好地适应了这一实际情况。

[12] For the production of a set of three voltages phase-displaced by 120 electrical degrees in time, a minimum of three coils phase-displaced 120 electrical degrees in space must be used.

为了产生三个时间相位相差120°的电压，最少需要三个空间上相差120°电角度的线圈。

[13] This flux reacts with the flux created by the field current, and electromechanical torque results from the tendency of these two magnetic fields to align.

这个磁通与励磁电流产生的磁通相互作用，机电转矩来源于这两个磁场试图对齐的趋势。

[14] In a synchronous motor, the steady-state speed is determined by the number of poles and the frequency of the armature current.

在一台同步电动机中，电动机极数和电枢电流的频率决定了稳态运行的速度。

[15] The flux produced by currents in the armature of a synchronous motor rotates ahead of that produced by the field, thus pulling on the field (and hence on the rotor) and doing work.

同步电动机的电枢磁场旋转并超前于励磁磁场，于是拖动励磁磁场（即转子）旋转并做功。

Chapter 14 Transformers

14.1 Introduction

Transformers are one of the most important components of any power system. It basically changes the level of voltages from one value to the other at constant frequency. Being a static machine the efficiency of a transformer could be as high as 99%.

Big generating stations are located at hundreds or more kilometres away from the load center (where the power will be actually consumed). Long transmission lines carry the power to the load centre from the generating stations. Generator is a rotating machines and the level of voltage at which it generates power is limited to several kilo volts only a typical value is 11 kV. To transmit large amount of power (several thousands of mega watts) at this voltage level means large amount of current has to flow through the transmission lines.[1] The cross sectional area of the conductor of the lines accordingly should be large. Hence cost involved in transmitting a given amount of power rises many folds. The transmission lines have their own resistances. This huge amount of current will cause tremendous amount of power loss or I^2r loss in the lines. This loss will simply heat the lines and becomes a wasteful energy. In other words, efficiency of transmission becomes poor and cost involved is high.

The above problems may be addressed if we could transmit power at a very high voltage, at 200 kV or 400 kV or even higher at 800 kV. But as pointed out earlier, a generator is incapable of generating voltage at these levels due to its own practical limitation. The solution to this problem is to use an appropriate step-up transformer at the generating station to bring the transmission voltage level at the desired value as depicted in Fig. 14-1 where for simplicity single phase system is shown to understand the basic idea.[2] Obviously when power reaches the load centre, one has to step down the voltage to suitable and safe values by using transformers. Thus transformers are an integral part in any modern power system. Transformers are located in places called substations. In cities or towns you must have noticed transformers are installed on poles—these are called pole mounted distribution transformers.

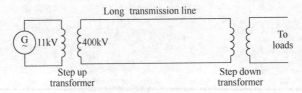

Fig. 14-1 A Simple Single Phase Power System

14.2 Transformer Construction

A transformer consists essentially of a laminated iron core linked with two windings of insulated wire. Its action depends upon the mutual induction which takes place between these two windings. Power is supplied to one of them at a definite frequency and voltage and is taken from the other at the same frequency but generally at a different voltage. The ratio of the two voltages depends upon the relative number of turns in the two windings. The winding to which power is supplied is called the primary; the other, which delivers power to the receiving circuit, is called the secondary. [3] Either will serve equally well as primary or as secondary. If the primary winding has more turns than the secondary winding, the voltage will be lowered and the transformer is called a step-down transformer. [4] If the secondary winding had the greater number of turns, the voltage will be raised and the transformer is called a step-up transformer.

There are two more or less distinct types of transformers which differ in the relative positions occupied by the windings and the iron core. These are the core and the shell types. The two types of transformers are shown in their simplest forms in Fig. 14-2. In the core type the windings envelop a considerable part of the magnetic circuit, while in the shell type the magnetic circuit envelops a considerable portion of the windings. [5] As a result of these differences, the core type of transformer, as compared with the shell type, has a core of small cross-section and long mean length and windings of a relatively great number of turns of small mean length. For a given output and voltage rating, the core type will contain less iron but more copper than the shell type. By proper design both types of transformers may be made to have essentially the same electrical characteristics, but when designed for approximately the same flux densities and current densities in the copper, the shell type of the two will have the larger iron loss and the smaller copper loss. [6] The almost universal use during the last few years of silicon steel sheet with its small iron loss for the cores of transformers has made design favor the shell type of transformer in the majority of cases. The shell type is the better for large transformers as it permits better bracing of the coils against displacements caused by short-circuits. [7] Under normal conditions the stresses between the windings and between successive turns of transformers are low, but at times of short-circuit they may be very great. A modern transformer may give from 25 to 50 times its full-load current on short-circuit if full voltage is maintained on its primary. Under such conditions the stresses between the windings would be from $(25)^2 = 625$ to $(50)^2 = 2500$, those at full load. The stresses on short-circuit are extremely important in large transformers. The core type works out best for very high voltages chiefly on account of greater space required for insulating the high- and low-tension coils of a shell-type transformer from one another. The space factor with the pan-cake type of coils used on shell transformers is poor and for very high-voltage transformers is often not over 0.3, while for the cylindrical coils used on the core type it may be, under similar conditions, as high as 0.4. Inherently the shell-type transformer has higher reactance than the core type of transformer on account of the type of coils used. Of the two types, the shell is the more expensive to repair.

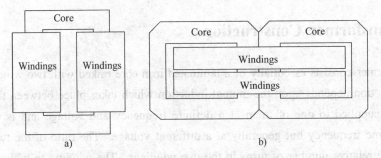

Fig. 14-2 Two Types of Transformers
a) The Core Type Transformer b) The Shell Type Transformer

14.2.1 Cores

1. Laminated Steel Cores (Fig. 14-3)

Transformers for use at power or audio frequencies typically have cores made of high permeability silicon steel. The steel has a permeability many times that of free space, and the core thus serves to greatly reduce the magnetizing current, and confine the flux to a path which closely couples the windings. [8] Early transformer developers soon realized that cores constructed from solid iron resulted in prohibitive eddy-current losses, and their designs mitigated this effect with cores consisting of bundles of insulated iron wires. Later designs constructed the core by stacking layers of thin steel laminations, a principle that has remained in use. Each lamination is insulated from its neighbors by a thin non-conducting layer of insulation. The universal transformer equation indicates a minimum cross-sectional area for the core to avoid saturation.

The effect of laminations is to confine eddy currents to highly elliptical paths that enclose little flux, and so reduce their magnitude. Thinner laminations reduce losses, but are more laborious and expensive to construct. Thin laminations are generally used on high frequency transformers, with some types of very thin steel laminations able to operate up to 10 kHz.

One common design of laminated core is made from interleaved stacks of E-shaped steel sheets capped with I-shaped pieces, leading to its name of "E-I transformer". Such a design tends to exhibit more losses, but is very economical to manufacture. The cut-core or C-core type is made by winding a steel strip around a rectangular form and then bonding the layers together. It is then cut in two, forming two C shapes, and the core assembled by binding the two C halves together with a steel strap. [9] They have the advantage that the flux is always oriented parallel to the metal grains, reducing reluctance.

A steel core's remanence means that it retains a static magnetic field when power is removed. When power is then reapplied, the residual field will cause a high inrush current until the effect of the remaining magnetism is reduced, usually after a few cycles of the applied alternating current. [10] Overcurrent protection devices such as fuses must be selected to allow this harmless inrush to pass. On transformers connected to long, overhead power transmission lines, induced currents due to geomagnetic disturbances during solar storms can cause saturation of the core and operation of transformer protection devices.

Fig. 14-3 Laminated Core Transformer Showing Edge of Laminations at Top of Unit

2. Solid Cores

Powdered iron cores are used in circuits (such as switch-mode power supplies) that operate above main frequencies and up to a few tens of kilohertz. These materials combine high magnetic permeability with high bulk electrical resistivity. For frequencies extending beyond the VHF band, cores made from non-conductive magnetic ceramic materials called ferrites are common. Some radio-frequency transformers also have movable cores (sometimes called 'slugs') which allow adjustment of the coupling coefficient (and bandwidth) of tuned radio-frequency circuits. [11]

3. Toroidal Cores (Fig. 14-4)

Toroidal transformers are built around a ring-shaped core, which, depending on operating frequency, is made from a long strip of silicon steel or permalloy wound into a coil, powdered iron, or ferrite. A strip construction ensures that the grain boundaries are optimally aligned, improving the transformer's efficiency by reducing the core's reluctance. The closed ring shape eliminates air gaps inherent in the construction of an E-I core. The cross-section of the ring is

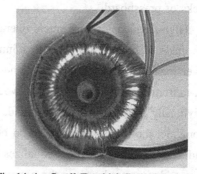

Fig. 14-4 Small Toroidal Core Transformer

usually square or rectangular, but more expensive cores with circular cross-sections are also available. The primary and secondary coils are often wound concentrically to cover the entire surface of the core. This minimizes the length of wire needed, and also provides screening to minimize the core's magnetic field from generating electromagnetic interference.

Toroidal transformers are more efficient than the cheaper laminated E-I types for a similar power level. Other advantages compared to E-I types, include smaller size (about half), lower weight (about half), less mechanical hum (making them superior in audio amplifiers), lower exterior magnetic field (about one tenth), low off-load losses (making them more efficient in standby circuits), single-bolt mounting, and greater choice of shapes. [12] The main disadvantages are higher cost and limited power capacity.

Ferrite toroidal cores are used at higher frequencies, typically between a few tens of kilohertz to a megahertz, to reduce losses, physical size, and weight of switch-mode power supplies. A drawback of toroidal transformer construction is the higher cost of windings. As a consequence, toroidal transformers are uncommon above ratings of a few kV · A. Small distribution transformers may achieve some of the benefits of a toroidal core by splitting it and forcing it open, then inserting a bobbin containing primary and secondary windings.

14.2.2 Windings (Fig. 14-5)

The conducting material used for the windings depends upon the application, but in all cases the individual turns must be electrically insulated from each other to ensure that the current travels throughout every turn. [13] For small power and signal transformers, in which currents are low and the potential difference between adjacent turns is small, the coils are often wound from enamelled magnet wire, such as Formvar wire. Larger power transformers operating at high voltages may be wound with copper rectangular strip conductors insulated by oil-impregnated paper and blocks of pressboard.

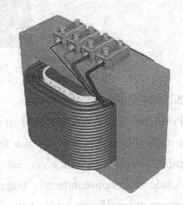

Fig. 14-5 Windings Are Usually Arranged Concentrically to Minimize Flux Leakage

High-frequency transformers operating in the tens to hundreds of kilohertz often have windings made of braided litz wire to minimize the skin-effect and proximity effect losses. [14] Large power transformers use multiple-stranded conductors as well, since even at low power frequencies non-uniform distribution of current would otherwise exist in high-current windings. Each strand is individually insulated, and the strands are arranged so that at certain points in the winding, or throughout the whole winding, each portion occupies different relative positions in the complete conductor. The transposition equalizes the current flowing in each strand of the conductor, and reduces eddy current losses in the winding itself. The stranded conductor is also more flexible than a solid conductor of similar size, aiding manufacture.

For signal transformers, the windings may be arranged in a way to minimize leakage inductance and stray capacitance to improve high-frequency response. [15] This can be done by splitting up each coil into sections, and those sections placed in layers between the sections of the other winding. This is known as a stacked type or interleaved winding.

Both the primary and secondary windings on power transformers may have external connections, called taps, to intermediate points on the winding to allow selection of the voltage ratio. The taps may be connected to an automatic on-load tap changer for voltage regulation of distribution circuits. Audio-frequency transformers, used for the distribution of audio to public address loudspeakers, have taps to allow adjustment of impedance to each speaker. A center-tapped transformer is often used in the output stage of an audio power amplifier in a push-pull circuit. Modulation transformers in AM transmitters are very similar.

Certain transformers have the windings protected by epoxy resin. By impregnating the transformer with epoxy under a vacuum, one can replace air spaces within the windings with epoxy, thus sealing the windings and helping to prevent the possible formation of corona and absorption of dirt or water. This produces transformers more suited to damp or dirty environments, but at increased manufacturing cost.

14.3 Ideal Transformers

To understand the working of a transformer it is always instructive, to begin with the concept of an ideal transformer with the following properties:

1) Primary and secondary windings have no resistance.
2) All the flux produced by the primary links the secondary winding i, e., there is no leakage flux.
3) Permeability μ_r of the core is infinitely large. In other words, to establish flux in the core vanishingly small (or zero) current is required.
4) Core loss comprising of eddy current and hysteresis losses are neglected.

Let us assume a sinusoidally varying voltage is impressed across the primary with secondary winding open circuited, as shown in Fig. 14-6. Although the current drawn I_m will be practically zero, but its position will be 90° lagging with respect to the supply voltage. The flux produced will obviously be in phase with I_m. In other words the supply voltage will lead the flux phasor by 90°. Since flux is common for both the primary and secondary coils, it is customary to take flux phasor as the reference.

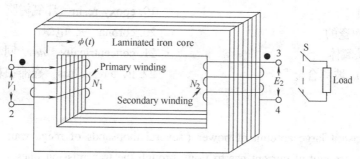

Fig. 14-6 A Typical Transformer

Let, $\phi(t) = \phi_{max} \sin \omega t$. Then,

$$v_1 = V_{max} \sin \left(\omega t + \frac{\pi}{2} \right) \quad (14\text{-}1)$$

The time varying flux $\phi(t)$ will link both the primary and secondary turns inducing in voltages e_1 and e_2 respectively. Instantaneous induced voltage e_1 and e_2 are given by

$$e_1 = -N_1 \frac{d\phi}{dt} = \omega N_1 \phi_{max} \sin \left(\omega t - \frac{\pi}{2} \right) = 2\pi f N_1 \phi_{max} \sin \left(\omega t - \frac{\pi}{2} \right)$$

$$e_2 = -N_2 \frac{d\phi}{dt} = \omega N_2 \phi_{max} \sin \left(\omega t - \frac{\pi}{2} \right) = 2\pi f N_2 \phi_{max} \sin \left(\omega t - \frac{\pi}{2} \right) \quad (14\text{-}2)$$

Magnitudes of the rms induced voltages will therefore be

$$E_1 = \sqrt{2}\pi f N_1 \phi_{max} = 4.44 f N_1 \phi_{max}$$
$$E_2 = \sqrt{2}\pi f N_2 \phi_{max} = 4.44 f N_2 \phi_{max} \tag{14-3}$$

The time phase relationship between the applied voltage v_1 and e_1 and e_2 will be same. The 180° phase relationship obtained in the mathematical expressions of the two merely indicates that the induced voltage opposes the applied voltage as Lenz's law. In other words if e_1 were allowed to act alone it would have delivered power in a direction opposite to that of v_1. By applying Kirchoff's law in the primary one can easily say that $V_1 = E_1$ as there is no other drop existing in this ideal transformer. Thus udder no load condition,

$$\frac{V_2}{V_1} = \frac{E_2}{E_1} = \frac{N_2}{N_1} \tag{14-4}$$

where, V_1, V_2 are the terminal voltages and E_1, E_2 are the rms induced voltages.

New Words and Expressions

1. step-up transformer   升压变压器
2. core type   铁心式
3. shell type   壳式
4. on account of   因为，由于
5. silicon steel sheet   硅钢片
6. interleave   *adj.* 交错的
7. steel strip   钢带
8. VHF   甚高频
9. ceramic   *n.* 陶瓷的
10. ferrite   *n.* 铁氧体
11. permalloy   *n.* 坡莫合金
12. powdered iron   铁粉
13. grain boundary   晶粒边界
14. bobbin   *n.* 线圈架
15. enamelled wire   漆包线
16. oil-impregnated   *adj.* 油浸的
17. pressboard   *n.* 压板
18. transposition   *n.* 变换
19. epoxy resin   环氧树脂
20. corona   *n.* 电晕
21. concentrically   *adv.* 同心地
22. hysteresis loss   磁滞损耗

Notes

[1] To transmit large amount of power (several thousands of mega watts) at this voltage level means large amount of current has to flow through the transmission lines.

为了在这个电压等级上传输大功率的电能（几千兆瓦）就意味着必须有强大的电流流过输电线。

[2] The solution to this problem is to use an appropriate step-up transformer at the generating station to bring the transmission voltage level at the desired value as depicted in Fig. 14-1 where for simplicity single phase system is shown to understand the basic idea.

关于这个问题的解决方案就是在发电站使用一个合适的升压变压器将输电电压提升到合适的值。为了说明这一方法的基本思想，图 14-1 所示描绘了一个简单的单相系统。

[3] The winding to which power is supplied is called the primary; the other, which delivers power to the receiving circuit, is called the secondary.

连接到供电电源一侧的绕组称作一次绕组，而另一个，输送电力给用电电路的绕组称作

二次绕组。

[4] If the primary winding has more turns than the secondary winding, the voltage will be lowered and the transformer is called a step-down transformer.

如果一次绕组的匝数多于二次绕组，电压会降低，而这个变压器因此称作降压变压器。

[5] In the core type the windings envelop a considerable part of the magnetic circuit, while in the shell type the magnetic circuit envelops a considerable portion of the windings.

在铁心式变压器中，绕组包住了磁路的一大部分，而在壳式变压器中磁路包住了绕组的一大部分。

[6] By proper design both types of transformers may be made to have essentially the same electrical characteristics, but when designed for approximately the same flux densities and current densities in the copper, the shell type of the two will have the larger iron loss and the smaller copper loss.

通过合理设计的两种形式的变压器本质上应该具有相同的电气特性，但是如果在设计中保持相同的磁密以及电密的话，壳式变压器会产生更高的铁耗以及更低的铜耗。

[7] The shell type is the better for large transformers as it permits better bracing of the coils against displacements caused by short-circuits.

对于大型变压器来说最好选择壳式变压器，因为在遭受短路时这种变压器的绕组更加牢靠。

[8] The steel has a permeability many times that of free space, and the core thus serves to greatly reduce the magnetizing current, and confine the flux to a path which closely couples the windings.

硅钢片的磁导率是空气的许多倍，因而，铁心能极大地减小磁化电流，并将磁通限制在连接绕组的闭合路径中。

[9] It is then cut in two, forming two C shapes, and the core assembled by binding the two C halves together with a steel strap.

然后这个方形铁心被切成两半，形成两个 C 形，再通过钢带将两个 C 形铁心捆绑在一起。

[10] When power is then reapplied, the residual field will cause a high inrush current until the effect of the remaining magnetism is reduced, usually after a few cycles of the applied alternating current.

当变压器再次接入电路时，剩磁会引起很大的励磁涌流直到剩磁的影响减小为止，通常这一过程要持续几个周期。

[11] Some radio-frequency transformers also have movable cores (sometimes called 'slugs') which allow adjustment of the coupling coefficient (and bandwidth) of tuned radio-frequency circuits.

有些射频变压器同样具有可移动的铁心（有时称作"蛞蝓"）用于射频调谐电路调整耦合系数。

[12] Other advantages compared to E-I types, include smaller size (about half), lower weight (about half), less mechanical hum (making them superior in audio amplifiers), lower ex-

terior magnetic field (about one tenth), low off-load losses (making them more efficient in stand-by circuits), single-bolt mounting, and greater choice of shapes.

相对 E-I 型变压器，环形变压器还有其他的优点：包括更小的尺寸（大约一半）、更低的重量（大约一半）、更低的机械噪声（适于音频放大器）、更低的外磁场（约 1/10）、低的空载损耗（在待机电路中更加高效）、单螺栓安装，以及在形状选择上更加自由。

[13] The conducting material used for the windings depends upon the application, but in all cases the individual turns must be electrically insulated from each other to ensure that the current travels throughout every turn.

制造绕组的材料取决于具体的应用，但是在所有应用中每匝线圈之间必须电气绝缘从而确保电流流过每一匝线圈。

[14] High-frequency transformers operating in the tens to hundreds of kilohertz often have windings made of braided litz wire to minimize the skin-effect and proximity effect losses.

工作在数十到数百赫兹的高频变压器绕组是由编织线构成的，用以使趋肤效应和邻近效应损耗最小化。

[15] For signal transformers, the windings may be arranged in a way to minimize leakage inductance and stray capacitance to improve high-frequency response.

对于信号变压器，绕组的构成方式应使漏电感和杂散电容最小化，以改善高频响应。

Chapter 15 Permanent Magnet Machines

15.1 Introduction

A permanent magnet (PM) machine is a machine where the excitation field is provided by a PM instead of a coil. The PM can be placed on the rotor or stator, in general, thus eliminating the requirement of a source of direct current for excitation. This results in a simple and rugged machine. The cross section of one type PM machine is shown in Fig. 15-1. The structure of stator core is the same as in a synchronous machine with an excitation coil. The rotor has a cylindrical steel core with radial or parallel magnetized PM on its surface or inside. The PM might be made of the neodymium material. In small, low cost machines, they might be made of ferrite magnetic material. PM motors supplied from inverters have become increasingly attractive for application in a wide range of speed applications, particularly following the introduction of Nd-FeB and SmCo magnet materials. [1] Most applications

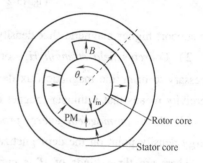

Fig. 15-1 Cross Section of a PM Machine

are for low- and medium-power levels but the range is continually being extended. PM motors can produce more steady-state and transient torques than induction machines of the same size. They also can give higher efficiency.

15.2 Introduction of PM Materials

The development of PM machine depends on the development of PM materials. A PM material is a kind of special material that is magnetized and produces persistent magnetic field without external energy. [2] According to the manufacturing process and component, the common PM material can be classified as shown in Fig. 15-2.

As any other ferromagnetic materials, a PM material can be described by the B-H hysteresis loop. PMs are also called hard magnetic materials, meaning ferromagnetic materials with a wide hysteresis loop. The basis for the evaluation of a PM is the portion of its hysteresis loop located in the upper left-hand quadrant, called the demagnetization curve. PM materials are characterized by the parameters listed below:

1) *Residual or remanent magnetic flux density B_r*, is the magnetic flux density corresponding to zero magnetic field intensity in the demagnetization curve. [3] High remanence means the magnet

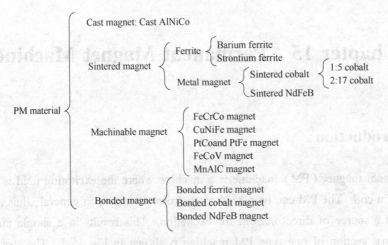

Fig. 15-2 Classification of PM Materials

can support higher magnetic flux density in the air gap of the magnetic circuit.

2) *Coercive field strength* H_c, or coercivity, is the value of demagnetizing field intensity necessary to bring the magnetic flux density to zero in a material previously magnetized.[4] High coercivity means that a thinner magnet can be used to withstand the demagnetization field.

3) *Maximum magnetic energy product* $(BH)_{max}$, is the product corresponds to the maximum energy density point on the demagnetization curve with coordinates B and H.

There are three kinds of PMs currently used for electric machines: AlNiCo, Ferrite, and Rare-earth material such as SmCo and NdFeB magnets. The corresponding demagnetization curves are given in Fig. 15-3.

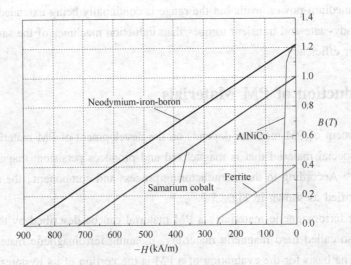

Fig. 15-3 Demagnetization Curves of Rare-Earth, Ferrite, and AlNiCo Magnets

The main advantages of AlNiCo are its high residual magnetic flux density and low temperature coefficient. The temperature coefficient of B_r is $-0.02\%/℃$ and maximum service temperature is 520℃. These advantages allow a high air gap magnetic flux density at high magnet temperature. Unfortunately, the coercive force is very low and the demagnetization curve is extremely non-linear. Therefore, it is very easy not only to magnetize but also to demagnetize AlNiCo. AlNiCo has been used in PM machines with relatively large air gaps. This results in a negligible armature reaction magnetic flux acting on the PMs.

Barium and strontium ferrites were invented in the 1950s. A ferrite has a higher coercive force than AlNiCo, but at the same time has a lower remanent magnetic flux density. The main advantages of ferrites are their low cost and very high electric resistance, which means no eddy-current losses in the magnet. Ferrite magnets are the most economical in fractional horsepower motors and may show an economic advantage over AlNiCo up to about 7.5kW. Ferrite magnets are commonly used in small motors for automobiles and electric toys.

Rare-earth magnets are strong PMs made from alloys of rare-earth elements. Developed in the 1970s and 1980s, rare-earth magnets are the strongest type of PMs, producing significantly stronger magnetic fields than other types such as ferrite or AlNiCo magnets. There are two types: neodymium magnets and samarium-cobalt magnets. The most important applicationfield of rare-earth magnets with about 40% of the sales is electrical machine. PM machines are used in a broad power range from a few mW to more than 1MW, covering a wide variety of applications from stepping motors for wristwatches via industrial servo drives for machine tools to large synchronous motors. High performance rare-earth magnets have successfully replaced AlNiCo and ferrite magnets in all applications where the high power density, improved dynamic performance or higher efficiency are of the prime interest.

15.3 Classification of PM Motors

In general, PM motors can be classified into:
1) DC commutator motors;
2) DC brushless motors;
3) AC synchronous motors.

The construction of a PM DC commutator motor is similar to a DC motor with the electromagnetic excitation system replaced by PMs. PM DC brushless and AC synchronous motor designs are practically the same: with a polyphase stator and PMs located on the rotor. The only difference is in the control and shape of the excitation voltage: an AC synchronous motor is fed with more or less sinusoidal waveforms which in turn produce a rotating magnetic field. In PM DC brushless motors the armature current has a shape of a square (trapezoidal) waveform, only two phase windings conduct the current at the same time and the switching pattern is synchronized with the rotor angular position.[5] The armature current of synchronous and DC brushless motors is not transmitted through brushes, which are subject to wear and require maintenance. Another advantage of the brushless motor is the fact that the power losses occur in the stator, where heat transfer conditions are good. Consequently the power density can be increased as compared with a DC commutator motor. In addition, considerable improvements in dynamics can be achieved because the air gap magnetic flux density is high, the rotor has a lower inertia and there are no speed-dependent current limitations. Thus, the volume of a brushless PM motor can be reduced by more than 40% while still keeping the same rating as that of a PM commutator motor.

15.4 Operational Principle of PM Motors

15.4.1 Operational Principle of PM Commutator Motors

In a DC motor, an armature rotates inside a magnetic field. Basic working principle of a DC motor is based on the fact that whenever a current carrying conductor is placed inside a magnetic field, there will be mechanical force experienced by that conductor. [6] All kinds of DC motors work in this principle only. Hence for constructing a DC motor it is essential to establish a magnetic field. The magnetic field is obviously established by means of magnet. The magnet can be any types i.e. it may be electromagnet or it can be PM. When PM is used to create magnetic field in a DC motor, the motor is referred as PMDC motor. Have you ever uncovered any battery operated toy? If you did, you had obviously found a battery operated motor inside it. This battery operated motor is nothing but a permanent magnet DC motor, shown in Fig. 15-4.

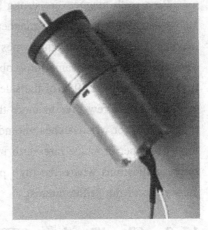

Fig. 15-4 Picture of a PMDC Motor

These types of motor are essentially simple in construction. These motors are commonly used as starter motors in automobiles, windshield wipers, washer, for blowers used in heaters and air conditioners, to raise and lower windows. It also extensively is used in toys. [7] As the magnetic field strength of a permanent magnet is fixed it cannot be controlled externally, field control of this type of DC motor is impossible. Thus PMDC motor is used where there is no need of speed control of motor by means of controlling its field. Small fractional and sub fractional kW motors now are constructed with PMs.

As it is indicated in the name of PMDC motor, the field poles of this motor are essentially made of PMs. A PMDC motor mainly consists of two parts, a stator and an armature. Here the stator is a steel cylinder. The magnets are mounted in the inner periphery of this cylinder. The magnets are mounted in such a way that the N-pole and S-pole of each magnet are alternatively faced towards armature. That means, if N-pole of one magnet is faced towards armature then S—pole of very next magnet is faced towards armature.

The rotor of a PMDC motor is similar to other DC motors. The rotor or armature of a PMDC motor also consists of core, windings and commutator. The armature core is made of a number of varnish insulated, slotted circular lamination of steel sheets. By fixing these circular steel sheets one by one, a cylindrical shaped slotted armature core is formed. The varnish insulated laminated steel sheets are used to reduce eddy current loss in the armature of a PMDC motor. These slots on the outer periphery of the armature core are used for housing armature conductors in them. The armature conductors are connected in a suitable manner which gives rise to armature winding. The end terminals of the winding are connected to the commutator segments placed on the motor shaft. Like common DC motor, carbon or graphite brushes are placed with spring pressure on the commutator segments to supply current to the armature.

As we said earlier the working principle of the PMDC motor is just similar to the general working principle of the DC motor. That is when a carrying conductor comes inside a magnetic field, a mechanical force will be experienced by the conductor and the direction of this force is governed by Fleming's left hand rule. As in a PMDC motor, the armature is placed inside the magnetic field of magnet; the armature rotates in the direction of the generated force. [8] Here each conductor of the armature experiences the mechanical force $F = BIL$ Newton where, B is the magnetic flux density in Tesla, I is the current in Ampere flowing through that conductor and L is the length of the conductor in meter comes under the magnetic field. Each conductor of the armature experiences a force and the summation of those forces produces a torque, which tends to rotate the armature.

15.4.2 Operational Principle of PM Brushless DC Motors

The availability of efficient semiconductor switches has provided means for eliminating the mechanical switching on commutator machines while retaining many useful properties. This kind of machine is called an electronically switched PM motor or a PM brushless DC (BLDC) motor. Fig. 15-5 shows a picture of a PM BLDC motor with controller. It is applied in four axis aircraft.

A PM BLDC motor consists of three parts, a motor, the rotor position sensors and electronic switching circuit. Principle block diagram of a PM BLDC motor is shown in Fig. 15-6.

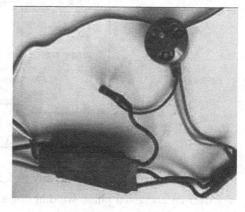

Fig. 15-5 Picture of a PM BLDC Motor

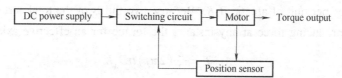

Fig. 15-6 Principle Block Diagram of a PM BLDC Motor

Fig. 15-7 shows a cross section of a two-pole motor. The rotor has two surface-mounted PMs, each covering approximately 180° of the rotor periphery and thereby producing a nearly rectangular space wave of flux density in the air gap. The stator has three phase windings but differs from the windings for three phase alternating machines in that the conductors of each phase are distributed approximately uniformly in slots over two arcs of 60° for each phase.

The motor can be supplied from a system such as that shown in Fig. 15-8. The motor phases are connected to a source of controllable direct current I through six electronic switches. At any instant, one upper and one lower switch are closed, connecting two phases in series to the current supply. [9] As seen in Fig. 15-7, each rotor magnet interacts with two 60° arcs of stator con-

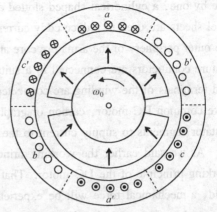

Fig. 15-7 Cross Section of a PM BLDC Motor

ductors carrying current with the polarity shown. When the rotor magnet edges reach the boundary between stator phases, a detector, such as a Hall device mounted on the stator, detects the reversal of the air-gap field and causes an appropriate opening and closing of the switches. [10] At the position shown in Fig. 15-1 and with the rotor rotating in a counterclockwise direction, switch 1 will open and switch 3 will close, thus energizing phases b and c and continuing the torque.

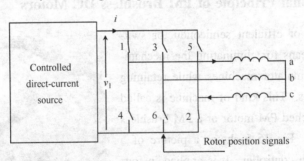

Fig. 15-8 Switching Circuit of a PM BLDC Motor

Suppose the air-gap flux density is B_g. If there are N_s turns per phase, the linear current density over an energized phase winding for a rotor of radius r is

$$K = \frac{3N_s i}{\pi r} \quad (16\text{-}1)$$

The tangential force per unit of area is $B_g K$. Because two thirds of the total surface area of the stator is effective in producing force at any instant, the torque for an effective axial length l is

$$\begin{aligned} T &= \frac{2}{3}(2\pi r) lr B_g K \\ &= 4rl B_g N_s i \\ &= ki \end{aligned} \quad (16\text{-}2)$$

Here k is called torque constant. Thus, the motor produces a torque directly proportional to its supply current. Reversal of the torque direction is usually achieved by appropriate change in the switching signals from the field detectors.

To determine the generated voltage produced by two phase windings in series, consider the machine to be rotating atan angular velocity ω_0. In each of the $4N_s$ series conductors, the electric field intensity will be vB_g where $v = r\omega_0$. Thus, the generated voltage is

$$e = 4N_s rlB_g \omega_0 = k\omega_0 \qquad (16\text{-}3)$$

If the current source in Fig. 15-8 is replaced by a controllable voltage source v_t, the motor will operate at a speed approximately proportional to the source voltage. By analogy with a commutator motor,

$$\omega_0 = \frac{v_t - Ri}{k} = \frac{v_t}{k} - \frac{RT}{k^2} \qquad (16\text{-}4)$$

where R is the resistance of two phases in series. Switched PM motors of the general type shown in Fig. 15-8 are used in a wide variety of drive applications such as in robots, machine tools, and disk drives.

15.4.3 Operational Principle of PM Synchronous Motors

PM synchronous motor is polyphase synchronous motor with PM rotors. Thus they are similar to the synchronous machines described in the previous section with the exception that the field windings are replaced by PM. [11]

Fig. 15-1 is a schematic diagram of a three-phase PM synchronous motor. In fact, the PM synchronous motor can be readily analyzed with techniques simply by assuming that the machine is excited by a field current of constant value, making sure to calculate the various machine inductances based on the effective permeability of the PM motor. [12]

Fig. 15-9 shows a picture of rotor and stator of a typical PM synchronous motor. A speed and position sensor mounted on the rotor shaft. This sensor is used for control of the motor. A number of techniques may be used for shaft-position sensing, including Hall-effect devices, light-emitting diodes and phototransistors in combination with a pulsed wheel, and inductance principles. [13]

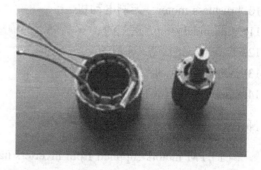

Fig. 15-9 Picture of a PM AC Synchronous Motor

A PM synchronous motor typically operates from the variable-frequency motor drive. Under conditions of constant-frequency, sinusoidal polyphase excitation, a PM synchronous motor behaves to a conventional AC synchronous motor with constant field excitation.

An alternate viewpoint of a PM synchronous motor is that it is a form of PM stepping motor with a non-salient stator. Under this viewpoint, the only difference between the two is that there will be little, if any, saliency torque in the PM synchronous motor. [14] In the simplest operation, the phases can be simply excited with stepped waveforms so as to cause the rotor to step sequentially from one equilibrium position to the next. Alternatively, using rotor-position feedback from a shaft-position sensor, the motor phase windings can be continuously excited in such a fashion as to control the torque and speed of the motor.[15] As with the stepping motor, the frequency of the excitation determines the motor speed, and the angular position between the rotor magnetic axis and a given phase and the level of excitation in that phase determines the torque which will be produced.

A PM synchronous motor is frequently referred to as brushless DC motors. This terminology comes about both because of the similarity, when combined with a variable-frequency, variable-voltage drive system, of their speed-torque characteristics to those of DC motors and because of the fact that one can view these motors as inside-out DC motors, with their field winding on the rotor and with their armature electronically commutated by the shaft-position sensor and by switches connected to the armature windings.

New Words and Expressions

1. magnetized   *adj.* 磁化的
2. neodymium   *n.* 钕
3. ferrite   *n.* 铁氧体
4. inverter   *n.* 变频器
5. demagnetization curve   退磁曲线
6. residual   *adj.* 剩余的
7. coercivity   *n.* 矫顽力
8. magnetic energy product   磁能积
9. rare-earth   *n.* 稀土
10. brushless motor   无刷电动机
11. sinusoidal   *adj.* 正弦的
12. trapezoidal   *adj.* 梯形的
13. maintenance   *n.* 维护
14. starter   *n.* 起动装置
15. field control   *n.* 励磁控制
16. eddy current loss   涡流损耗
17. carrying conductor   载流导体
18. semiconductor   *n.* 半导体
19. position sensor   位置传感器
20. surface-mounted   *adj.* 表面贴装的
21. detector   *n.* 探测器
22. counterclockwise   *n.* 逆时针
23. shaft   *n.* 轴
24. light-emitting diode   发光二极管
25. phototransistor   *n.* 光电晶体管
26. variable-frequency   *n.* 变频
27. stepping motor   步进电动机
28. non-salient stator   非凸极定子

Notes

[1] PM motors supplied from inverters have become increasingly attractive for application in a wide range of speed applications, particularly following the introduction of NdFeB and SmCo magnet materials.

变频器供电的永磁电动机在需要进行宽范围调速的场合应用越来越广泛，特别在钕铁硼磁体和钐钴磁体问世以来更是这样。

[2] A PM material is a kind of special material that is magnetized and produces persistent magnetic field without external energy.

永磁材料是一种能够在充磁之后不需要外部能量供应即可产生持续磁场的特殊材料。

[3] Residual or remanent magnetic flux density B_r, is the magnetic flux density corresponding to zero magnetic field intensity in the demagnetization curve.

剩余磁通密度是在退磁曲线上磁场强度为零时的磁通密度。

[4] Coercive field strength H_c, or coercivity, is the value of demagnetizing field intensity necessary to bring the magnetic flux density to zero in a material previously magnetized.

矫顽力是材料在预先磁化的情况下使得磁通密度等于零时所施加的退磁磁场的值。

[5] In PM DC brushless motors the armature current has a shape of a square (trapezoidal) waveform, only two phase winding conduct the current at the same time and the switching pattern is synchronized with the rotor angular position.

在永磁无刷电动机中，电枢电流形状为方波（梯形波），在任一瞬时仅由两相绕组导通并且开关模式与转子角位置保持同步。

[6] Basic working principle of a DC motor is based on the fact that whenever a current carrying conductor is placed inside a magnetic field, there will be mechanical force experienced by that conductor.

直流电动机的基本工作原理是基于这样的事实：放置在磁场中的载流导体受到机械力的作用。

[7] These motors are commonly used as starter motors in automobiles, windshield wipers, washer, for blowers used in heaters and air conditioners, to raise and lower windows. It also extensively is used in toys.

这些电动机（是指永磁直流电动机）通常应用于汽车的起动装置、风窗玻璃刮水器、洗涤器、加热器与空调器的风机、电动车窗，也大量应用于电动玩具中。

[8] As in a PMDC motor, the armature is placed inside the magnetic field of magnet; the armature rotates in the direction of the generated force.

在永磁直流电动机中，电枢放置在永磁体产生的磁场内部，电枢沿受力方向旋转。

[9] At any instant, one upper and one lower switch are closed, connecting two phases in series to the current supply.

在任一瞬时，三相桥中一个上臂和一个下臂开关闭合，从而使两相绕组与供电电源串联连接。

[10] When the rotor magnet edges reach the boundary between stator phases, a detector, such as a Hall device mounted on the stator, detects the reversal of the air-gap field and causes an appropriate opening and closing of the switches.

当转子永磁体转到定子某相边界时，探测器，如贴装在定子上的霍尔器件，探测到气隙磁场的反转并引起开关以一定顺序开断和闭合。

[11] Thus they are similar to the synchronous machines described in the previous section with the exception that the field windings are replaced by PM.

于是它们与前面单元所描述的同步电机类似，只是励磁绕组由永磁体取代罢了。

[12] In fact, the PM synchronous motor can be readily analyzed with techniques simply by assuming that the machine is excited by a field current of constant value, making sure to calculate the various machine inductances based on the effective permeability of the PM motor.

事实上，在分析永磁电机时仅需假设电机由一个恒值电流励磁即可，确保基于永磁电机有效磁导率来计算电机的各种电感。

[13] A number of techniques may be used for shaft-position sensing, including Hall-effect devices, light-emitting diodes and phototransistors in combination with a pulsed wheel, and inductance principles.

大量的技术用于转轴位置检测，这些技术包括霍尔器件，发光二极管与光电晶体管与脉冲盘组合，以及电感原理。

[14] An alternate viewpoint of a PM synchronous motor is that it is a form of PM stepping motor with a non-salient stator. Under this viewpoint, the only difference between the two is that there will be little, if any, saliency torque in the PM synchronous motor.

对于同步电动机一个替代的观点是它是带有非凸极定子的永磁步进电动机的一种形式。基于此观点，这两种电动机的唯一区别是永磁同步电动机的凸极扭矩即使有的话也是很少。

[15] Alternatively, using rotor-position feedback from a shaft-position sensor, the motor phase windings can be continuously excited in such a fashion as to control the torque and speed of the motor.

换句话说，利用取自轴位置传感器的转子位置反馈，电动机的相位绕组能像控制电动机的扭矩和速度常用的方式一样被持续地激励。

PART 4   POWER SYSTEMS

Chapter 16   Operating Characteristics of Modern Power Systems

16.1   Transmission and Distribution Systems

Electric power systems vary in size and structural components. However, they all have the same basic characteristics:

1) Being comprised of three-phase AC systems operating essentially at constant voltage. Generation and transmission facilities use three-phase equipment. Industrial loads are invariably three-phase. Single-phase residential and commercial loads are distributed equally among the phases so as to effectively form a balanced three-phase system. [1]

2) Using synchronous machines for generation of electricity. Prime movers convert the primary sources of energy (fossil, nuclear, and hydraulic) to mechanical energy that is, in turn, converted to electrical energy by synchronous generators.

3) Transmitting power over significant distances to consumers spread over a wide area. This requires a transmission system comprising subsystems operating at different voltage levels.

Electric power is produced at generating stations (GS) and transmitted to consumers through a complex network of individual components, including transmission lines, transformers, and switching devices.

The sources of electric power are usually interconnected by a transmission system or network that distributes the power to the various load point or load centers. A small portion of a transmission system that suggests the interconnections is shown as a one-line diagram in Fig. 16-1. Various symbols for generators, transformers, circuit breakers, loads, and the points of connection (nodes), called buses, are identified in the figure.

The generator voltages are in the range of 11kV to 30kV; higher generator voltages are difficult to obtain because of insulation problems in the narrow confines of the generator stator. Transformers are then used to step up the voltages to the range of 110kV to 765kV. In California the backbone of the transmission system is composed mainly of 500kV, 345kV, and 230kV three-phase lines. The voltages refer to voltages from line to line.

One reason for using high transmission-line voltages is to improve energy transmission efficiency. Basically, transmission of a given amount of power (at specified power factor) requires a fixed product of voltage and line current. Thus, the higher the voltage, the lower the current can be. Lower line currents are associated with lower resistive losses (I^2R) in the line. Another reason for higher voltages is the enhancement of stability.

A comment is in order about the loads shown in Fig. 16-1. The loads referred to here represent bulk load, such as the distribution system of a town, city, or large industrial plant. Such distribution systems provide power at various voltage levels. Large industrial consumers or railroads might accept power directly at voltage levels of 23kV to 138kV; they would then step down the voltages further. Smaller industrial or commercial consumers typically accept power at voltage levels of 4.16kV to 34.5kV. Residential consumers normally receive single-phase power from pole-mounted distribution transformers at voltage levels of 120/240V.

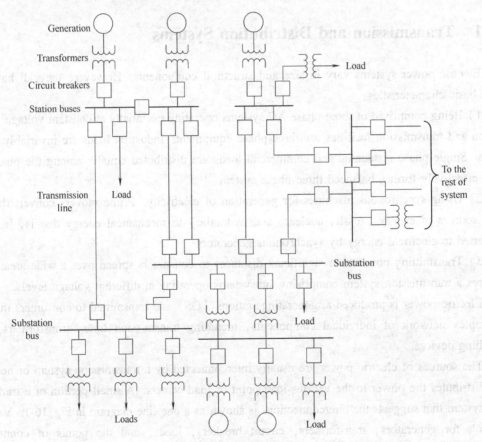

Fig. 16-1 One-Line Diagram

Although the transmission-distribution system is actually one interconnected system, it is convenient to separate out the transmission system, as we have done in Fig. 16-1. A similar diagram for the distribution system can be drawn with bulk substations replacing the generators as the sources of power and with lower-level loads replacing the bulk power loads shown in Fig. 16-1. [2]

16.2 Power System Controls

Automatic control systems are used extensively in power systems. Local controls are employed at turbine-generator units and at selected voltage-controlled buses. Central controls are employed at area control centers.

Fig. 16-2 shows two basic controls of a steam turbine-generator: the voltage regulator and turbine-governor. The voltage regulator adjusts the power output of the generator exciter in order to control the magnitude of generator terminal voltage V_t. When a reference voltage V_{ref} is raised (or lowered), the output voltage V_r of the regulator increases (or decreases) the exciter voltage E_{fd} applied to the generator field winding, which in turn acts to increase (or decrease) V_t.[3] Also a voltage transformer and rectifier monitor V_t, which is used as a feedback signal in the voltage regulator. If V_t decreases, the voltage regulator increases V_r to increase E_{fd}, which in turn acts to increase V_t.

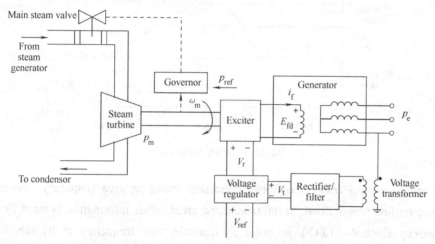

Fig. 16-2 Voltage Regulator and Turbine-Governor Controls for a Steam-Turbine Generator

The turbine-governor shown in Fig. 16-2 adjusts the steam valve position to control the mechanical power output p_m of the turbine. When a reference power level p_{ref} is raised (or lowered), the governor moves the steam valve in the open (or close) direction to increase (or decrease) p_m. The governor also monitors rotor speed ω_m, which is used as a feedback signal to control the balance between p_m and the electrical power output p_e of the generator.[4] Neglecting losses, if p_m is greater than p_e, ω_m increases, and the governor moves the steam valve in the close direction to reduce p_m. Similarly, if p_m is less than p_e, ω_m decreases, and the governor moves the valve in the open direction.

In addition to voltage regulators at generator buses, equipment is used to control voltage magnitudes at other selected buses. Tap-changing transformers, switched capacitor banks, and static var systems can be automatically regulated for rapid voltage control.

Central controls also play an important role in modern power systems. Today's systems are composed of interconnected areas, where each area has its own control center.[5] There are many advantages to interconnections. For example, interconnected areas can share their reserve power to handle anticipated load peaks and unanticipated generator outages. Interconnected areas can also tolerate larger load changes with smaller frequency deviations than an isolated area.

Fig. 16-3 shows how a typical area meets its daily load cycle. The base load is carried by base-loaded generators running at 100% of their rating for 24 hours. Nuclear units and large fossil-fuel units are typically base-loaded. The variable part of the load is carried by units that are controlled from the central control center. Medium-sized fossil-fuel units and hydro units are used for control. During peak load hours, smaller, less efficient units such as gas-turbine or diesel-generating units are employed. In addition, generators operating at partial output (with *spinning reserve*) and standby generators provide a reserve margin.

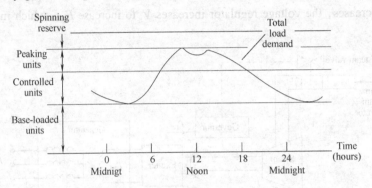

Fig. 16-3 Daily Load Cycle

The central control center monitors information including area frequency, generating unit outputs, and tie-line power flows to interconnected areas. This information is used by automatic Load-Frequency Control (LFC) in order to maintain area frequency at its scheduled value (60Hz) and net tie-line power flow out of the area at its scheduled value.[6] Raise and lower reference power signals are dispatched to the turbine-governors of controlled units.

Operating costs vary widely among controlled units. Larger units tend to be more efficient, but the varying cost of different fuels such as coal, oil, and gas is an important factor. Economic dispatch determines the megawatt outputs of the controlled units that minimize the total operating cost for a given load demand. Economic dispatch is coordinated with LFC such that reference power signals dispatched to controlled units move the units toward their economic loadings and satisfy LFC objectives. Optimal power flow combines economic dispatch with power flow so as to optimize generation without exceeding limits on transmission line load-ability.

16.3 Generator-Voltage Control

The exciter delivers DC power to the field winding on the rotor of a synchronous generator.

For older generators, the exciter consists of a DC generator driven by the rotor. The DC power is transferred to the rotor via slip rings and brushes. For newer generators, static or brushless exciters are often employed.

For static exciters, AC power is obtained directly from the generator terminals or a nearby station service bus. The AC power is then rectified via thyristors and transferred to the rotor of the synchronous generator via slip rings and brushes.

For brushless exciters, AC power is obtained from an "inverted" synchronous generator whose three-phase armature windings are located on the main generator rotor and whose field winding is located on the stator. [7] The AC power from the armature windings is rectified via diodes mounted on the rotor and is transferred directly to the field winding. For this design, slip rings and brushes are eliminated.

Block diagrams of several standard types of generator-voltage control systems have been developed by the IEEE Working Group on Exciters. A simplified block diagram of generator-voltage control is shown in Fig. 16-4. Nonlinearities due to exciter saturation and limits on exciter output are not shown in this figure.

The generator terminal voltage V_t in Fig. 16-4 is compared with a voltage reference V_{ref} to obtain a voltage error signal ΔV, which in turn is applied to the voltage regulator. The $1/(T_r s + 1)$ block accounts for voltage-regulator time delay, where s is the Laplace operator and T_r is the voltage-regulator time constant. Note that if a unit step is applied to a $1/(T_r s + 1)$ block, the output rises exponentially to unity with time constant T_r.

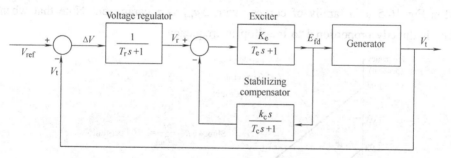

Fig. 16-4 Simplified Block Diagram—Generator-Voltage Control

The stabilizing compensator shown in Fig. 16-4 is used to improve the dynamic response of the exciter by reducing excessive overshoot. The compensator is represented by a $K_c s/(T_c s + 1)$ block, which provides a filtered first derivative. The input to this block is the exciter voltage E_{fd}, and the output is a stabilizing feedback signal that is subtracted from the regulator voltage V_r.

Block diagrams such as those shown in Fig. 16-4 are used for computer representation of generator-voltage control in transient stability computer programs. In practice, high-gain, fast-responding exciters provide large, rapid increases in field voltage E_{fd} during short circuits at the generator terminals in order to improve transient stability after fault clearing. [8] Equations represented in the block diagram can be used to compute the transient response of generator-voltage control.

16.4 Turbine-Governor Control

Turbine-generator units operating in a power system contain stored kinetic energy due to their rotating masses. If the system load suddenly increases, stored kinetic energy is released to initially supply the load increase. Also, the electrical torque T_e of each turbine-generating unit increases to supply the load increase, while the mechanical torque T_m of the turbine initially remains constant. From Newton's second law, $J\alpha = T_m - T_e$, the acceleration α is therefore negative. That is, each turbine-generator decelerates and the rotor speed drops as kinetic energy is released to supply the load increase. The electrical frequency of each generator, which is proportional to rotor speed for synchronous machines, also drops.

From this, we conclude that either rotor speed or generator frequency indicates a balance or imbalance of generator electrical torque T_e and turbine mechanical torque T_m. If speed or frequency is decreasing, then T_e is greater than T_m (neglecting generator losses). Similarly, if speed or frequency is increasing, T_e is less than T_m. Accordingly, generator frequency is an appropriate control signal for governing the mechanical output power of the turbine.

The steady-state frequency-power relation for turbine-governor control is

$$\Delta p_m = \Delta p_{ref} - \Delta f / R \quad (16-1)$$

where Δf is the change in frequency, Δp_m is the change in turbine mechanical power output, and Δp_{ref} is the change in a reference power setting. R is called the regulation constant. The equation is plotted in Fig. 16-5 as a family of curves, with Δp_{ref} as a parameter. Note that when Δp_{ref} is fixed, Δp_m is directly proportional to the drop in frequency.

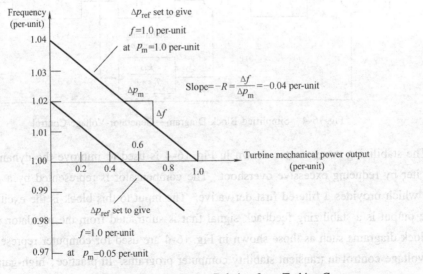

Fig. 16-5 Steady-State Frequency-Power Relation for a Turbine-Governor

Fig. 16-5 illustrates a steady-state frequency-power relation. When an electrical load change occurs, the turbine-generator rotor accelerates or decelerates, and frequency undergoes a transient

Chapter 16 Operating Characteristics of Modern Power Systems

disturbance.[9] Under normal operating conditions, the rotor acceleration eventually becomes zero, and the frequency reaches a new steady-state, shown in Fig. 16-5.

The regulation constant R in Equation (16-1) is the negative of the slope of the Δf versus Δp_m curves shown in Fig. 16-5. The units of R are Hz/MW when Δf is in Hz and Δp_m is in MW. When Δf and Δp_m are given in per-unit, however, R is also in per-unit.

The steady-state frequency-power relation for one area of an interconnected power system can be determined by summing Equation (16-1) for each turbine-generating unit in the area.

$$\Delta p_m = \Delta p_{m1} + \Delta p_{m2} + \Delta p_{m3} + \cdots$$

$$= (\Delta p_{ref1} + \Delta p_{ref2} + \cdots) - \left(\frac{1}{R_1} + \frac{1}{R_2} + \cdots\right)\Delta f \quad (16\text{-}2)$$

$$= \Delta p_{ref} - \left(\frac{1}{R_1} + \frac{1}{R_2} + \cdots\right)\Delta f$$

Noting that Δf is the same for each unit, where Δp_m is the total change in turbine mechanical powers and Δp_{ref} is the total change in reference power settings within the area. We define the area frequency response characteristic β as

$$\beta = \left(\frac{1}{R_1} + \frac{1}{R_2} + \cdots\right) \quad (16\text{-}3)$$

Using Equation (16-3) in Equation (16-2),

$$\Delta p_m = \Delta p_{ref} - \beta \Delta f \quad (16\text{-}4)$$

Equation (16-4) is the area steady-state frequency-power relation. The units of β are MW/Hz when Δf is in Hz and Δp_m is in MW. β can also be given in per-unit. In practice, β is somewhat higher than that given by Equation (16-3) due to system losses and the frequency dependence of loads.

A standard figure for the regulation constant is $R = 0.05$ per-unit. When all turbine-generating units have the same per-unit value of R based on their own ratings, then each unit shares total power changes in proportion to its own ratings.

Fig. 16-6 shows a block diagram of a non-reheat steam turbine-governor, which includes nonlinearities and time delays that were not included in Equation (16-1). The deadband block in this figure accounts for the fact that speed governors do not respond to change in frequency or to reference power settings that are smaller than a specified value. The limiter block accounts for the fact that turbines have minimum and maximum outputs. The $1/(Ts+1)$ blocks account for time delays, where s is the Laplace operator and T is a time constant. Typical values are $T_0 = 0.10$ s and

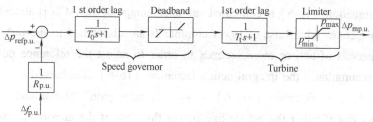

Fig. 16-6 Turbine-Governor Block Diagram

$T_t = 1.0$ s. Block diagrams for steam turbine-governors with reheat and hydro turbine-governors are also available.

16.5 Load-Frequency Control

As shown in section 16.4, turbine-governor control eliminates rotor accelerations and decelerations following load changes during normal operation. However, there is a steady-state frequency error Δf when the change in turbine-governor reference setting Δp_{ref} is zero. One of the objectives of Load-Frequency Control (LFC), therefore, is to return Δf to zero.

In a power system consisting of interconnected areas, each area agrees to export or import a scheduled amount of power through transmission-line interconnections, or tie-lines, to its neighboring areas.[10] Thus, a second LFC objective is to have each area absorb its own load change during normal operation. This objective is achieved by maintaining the net tie-line power flow out of each area at its scheduled value.

The following summarizes the two basic LFC objectives for an interconnected power system:

1) Following a load change, each area should assist in returning the steady-state frequency error Δf to zero.

2) Each area should maintain the net tie-line power flow out of the area at its scheduled value, in order for the area to absorb its own load changes.[11]

The following control strategy developed by N. Cohn meets these LFC objectives. We first define the Area Control Error (ACE) as follows:

$$ACE = (p_{tie} - p_{tie,sched}) + B_f(f - 60) \qquad (16-5)$$
$$= \Delta p_{tie} + B_f \Delta f$$

where Δp_{tie} is the deviation in net tie-line power flow out of the area from its scheduled value $\Delta p_{tie,sched}$, and Δf is the deviation of area frequency from its scheduled value (60Hz). Thus, the ACE for each area consists of a linear combination of tie-line error Δp_{tie} and frequency error Δf. The constant B_f is called a frequency bias constant.

The change in reference power setting Δp_{tie} of each turbine-governor operating under LFC is proportional to the integral of the area control error. That is,

$$\Delta p_{refi} = -K_i \int ACE dt \qquad (16-6)$$

Each area monitors its own tie-line power flows and frequency at the area control center. The ACE given by Equation (16-5) is computed and a percentage of the ACE is allocated to each controlled turbine-generator unit. Raise or lower commands are dispatched to the turbine-governors at discrete time intervals of two or more seconds in order to adjust the reference power settings. As the commands accumulate, the integral action Equation (16-6) is achieved.

The constant K_i in Equation (16-6) is an integrator gain. The minus sign in Equation (16-6) indicates that if either the net tie-line power flow out of the area or the area frequency is low—that is, if the ACE is negative—then the area should increase its generation.

When a load change occurs in any area, a new steady-state operation can be obtained only after the power output of every turbine-generating unit in the interconnected system reaches a constant value.[12] This occurs only when all reference power settings are zero, which in turn occurs only when the ACE of every area is zero. Furthermore, the ACE is zero in every area only when both Δp_{tie} and Δf are zero. Therefore, in steady-state, both LFC objectives are satisfied.

The choice of the B_f and K_i constants in Equation (16-5) and Equation (16-6) affects the transient response to load changes—for example, the speed and stability of the response. The frequency bias B_f should be high enough such that each area adequately contributes to frequency control. Cohn has shown that choosing B_f equal to the area frequency response characteristic, $B_f = \beta$, gives satisfactory performance of the interconnected system. The integrator gain K_i should not be too high; otherwise, instability may result. Also, the time interval at which LFC signals are dispatched, 2 or more seconds, should be long enough so that LFC does not attempt to follow random or spurious load changes. A detailed investigation of the effect of B_f, K_i and LFC time interval on the transient response of LFC and turbine-governor controls is beyond the scope of this text.

Two additional LFC objectives are to return the integral of frequency error and the integral of net tie-line error to zero in steady-state. By meeting these objectives, LFC controls both the time of clocks that are driven by 60Hz motors and energy transfers out of each area. These two objectives are achieved by making temporary changes in the frequency schedule and tie-line schedule in Equation (16-5).

Finally, note that LFC maintains control during normal changes in load and frequency—that is, changes that are not too large. During emergencies, when large imbalances between generation and load occur, LFC is bypassed and other emergency controls are applied.

16.6 Optimal Power Flow

The economic dispatch problem is that how the real power output of each controlled generating unit in an area is selected to meet a given load and to minimize the total operating costs in the area. Economic dispatch has one significant shortcoming—it ignores the limits imposed by the devices in the transmission system. Each transmission line and transformer has a limit on the amount of power that can be transmitted through it, with the limits arising because of thermal, voltage, or stability considerations. Traditionally, the transmission system was designed so that when the generation was dispatched economically there would be no limit violations. Hence, just solving economic dispatch was usually sufficient. However, with the worldwide trend toward deregulation of the electric utility industry, the transmission system is becoming increasingly constrained. For example, in the PJM power market in the eastern United States, the number of hours with active transmission line limit violations increased from 294 hours during the summer of 1998 to 548 hours during the summer of 1999.[13]

The solution to the problem of optimizing the generation while enforcing the transmission lines is to combine economic dispatch with the power flow. The result is known as the Optimal

Power Flow (OPF). There are several methods for solving the OPF, with the Linear Programming (LP) approach (this is the technique used with Power-World Simulator). The LP OPF solution algorithm iterates between solving the power flow to determine the flow of power in the system devices and solving an LP to economically dispatch the generation (and possibility over controls) subject to the transmission system limits. In the absence of system limits, the OPF generation dispatch will be identical to the economic dispatch solution.

16.7 Power System Stability

Power system stability refers to the ability of synchronous machines to move from one steady-state operating point following a disturbance to another steady-state operating point, without losing synchronism. There are three types of power system stability: steady-state, transient, and dynamic.

Steady-state stability, involves slow or gradual changes in operating points. Steady-state stability studies, which are usually performed with a power-flow computer program, ensure that phase angles across transmission lines are not too large, that bus voltages are close to nominal values, and that generators, transmission lines, transformers, and other equipment are not overloaded.

Transient stability involves major disturbances such as loss of generation, line-switching operations, faults, and sudden load changes. Following a disturbance, synchronous machine frequencies undergo transient deviations from synchronous frequency (60Hz), and machine power angles change. The objective of a transient stability study is to determine whether or not the machines will return to synchronous frequency with new steady-state power angles.[14] Changes in power flows and bus voltages are also of concern.

Legend gives an interesting mechanical analogy to the power system transient stability program. As shown in Fig. 16-7, a number of masses representing synchronous machines are interconnected by a network of elastic strings representing transmission lines. Assume that this network is initially at rest in steady-state, with the net force on each string below its break point, when one of the strings is cut, representing the loss of a transmission line.

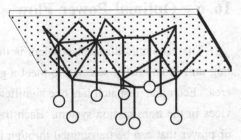

Fig. 16-7 Mechanical Analog of Power System Transient Stability

As a result, the masses undergo transient oscillations and the forces on the strings fluctuate. The system will then either settle down to a new steady-state operating point with a new set of string forces, or additional strings will break, resulting in an even weaker network and eventual system collapse. That is, for a given disturbance, the system is either transiently stable or unstable.

In today's large-scale power systems with many synchronous machines interconnected by complicated transmission networks, transient stability studies are best performed with a digital computer program.[15] For a specified disturbance, the program alternately solves, step by step, algebraic power-flow equations representing a network and nonlinear differential equations representing synchronous machines. Both pre-disturbance, disturbance, and post-disturbance computations are performed. The program output includes power angles and frequencies of synchronous machines, bus voltages, and power flows versus time.

In many cases, transient stability is determined during the first swing of machine power angles following a disturbance. During the first swing, which typically lasts about 1 second, the mechanical output power and the internal voltage of a generating unit are often assumed constant. However, where multiswings lasting several seconds are of concern, models of turbine-governors and excitation systems, as well as more detailed machine models can be employed to obtain accurate transient stability results over the longer time period.[16]

Dynamic stability involves an even longer time period, typically several minutes. It is possible for controls to affect dynamic stability even though transient stability is maintained. The action of turbine-governors, excitation systems, tap-changing transformers, and controls from a power system dispatch center can interact to stabilize or destabilize a power system several minutes after a disturbance has occurred.

New Words and Expressions

1. synchronous generator   同步发电机
2. prime mover   原动机
3. hydraulic   *adj.* 水力的
4. backbone   *n.* 主干网
5. power factor   功率因数
6. resistive   *adj.* 电阻的
7. turbine-generator   涡轮发电机
8. regulator   *n.* 调节器
9. governor   *n.* 调速器
10. generator exciter   发电机励磁器
11. field winding   励磁绕组
12. voltage transformer   电压互感器
13. rectifier   *n.* 整流
14. slip ring   集电环
15. brushless exciter   无刷励磁
16. thyristor   *n.* 晶闸管
17. armature winding   电枢绕组
18. electrical torque   电磁转矩
19. assist in   辅助，协助
20. bias   *n.* 偏差，偏移
21. integral   *adj.* 积分的
22. discrete   *adj.* 离散的
23. circuit breaker   断路器
24. substation   *n.* 变电站
25. tap-changing transformer   抽头变压器
26. outage   *n.* 断电
27. frequency deviation   频率偏移
28. diesel   *n.* 柴油
29. spinning reserve   热备用
30. optimal power flow   最优潮流
31. saturation   *n.* 饱和
32. unit step   单位阶跃
33. exponentially   *adv.* 以指数方式
34. compensator   *n.* 补偿装置
35. overshoot   *n.* 超调量
36. first derivative   一阶导数
37. deadband   *n.* 死区
38. bypass   *v.* 绕过，避开
39. phase angle   相位角
40. differential   *adj.* 微分的

Notes

[1] Single-phase residential and commercial loads are distributed equally among the phases so as to effectively form a balanced three-phase system.

单相住宅和商业负荷均衡地分布在各相中，因而有效地组成了三相平衡系统。

[2] A similar diagram for the distribution system can be drawn with bulk substations replacing the generators as the sources of power and with lower-level loads replacing the bulk power loads shown in Fig. 16-1.

一个相似的配电系统图可以由多个变电所和低压负荷组成，变电所代替了作为功率来源的发电机，而低压负荷代替了在图16-1中的功率负荷。

[3] When a reference voltage V_{ref} is raised (or lowered), the output voltage V_r of the regulator increases (or decreases) the exciter voltage E_{fd} applied to the generator field winding, which in turn acts to increase (or decrease) V_t.

当参考电压V_{ref}升高（或降低）时，调节器的输出电压V_r使施加在发电机励磁绕组的电压E_{fd}增加（或减少），随后发电机励磁绕组的励磁电压E_{fd}作用，从而增加（或减少）发电机端电压V_t。

[4] The governor also monitors rotor speed ω_m, which is used as a feedback signal to control the balance between p_m and the electrical power output p_e of the generator.

调节器也会监控转子速度ω_m，它被用作控制p_m和发电机的电功率输出信号p_e之间的平衡的一个反馈信号。

[5] Today's systems are composed of interconnected areas, where each area has its own control center.

现在的系统都是由互联区域组成的，其每个区域都有它自己的控制系统。

[6] This information is used by automatic Load-Frequency Control (LFC) in order to maintain area frequency at its scheduled value (60Hz) and net tie-line power flow out of the area at its scheduled value.

自动负荷-频率控制使用这些信息把频率维持在60Hz，并使流出该区域的净联络线功率为预定值。

[7] For brushless exciters, AC power is obtained from an "inverted" synchronous generator whose three-phase armature windings are located on the main generator rotor and whose field winding is located on the stator.

就无刷励磁机而言，需要从同步发电机中获取交流功率，其三相电枢绕组位于主发电机的转子上，其励磁绕组位于定子上。

[8] In practice, high-gain, fast-responding exciters provide large, rapid increases in field voltage E_{fd} during short circuits at the generator terminals in order to improve transient stability after fault clearing.

在实际中，高增益、快速响应的励磁器在发电机端短路时可以提供迅速增长的励磁电压E_{fd}，以提高故障清除后的暂态稳定。

[9] When an electrical load change occurs, the tuibine-generator rotor accelerates or decel-

erates, and frequency undergoes a transient disturbance.

当电力负荷发生变化时，该涡轮发电机的转子就会加速或减速，其频率就会出现一个短暂的干扰。

[10] In a power system consisting of interconnected areas, each area agrees to export or import a scheduled amount of power through transmission-line interconnections, or tie-lines, to its neighboring areas.

在一个由互联区域组成的电力系统中，通过互联传输线或者联络线，每一个区域都能够输入或输出既定功率给其相邻的区域。

[11] Each area should maintain the net tie-line power flow out of the area at its scheduled value, in order for the area to absorb its own load changes.

每一个区域应该按其预定值维持输出到其他区域的联络线功率流，以便吸收自己的负荷变化。

[12] When a load change occurs in any area, a new steady-state operation can be obtained only after the power output of every turbine-generating unit in the interconnected system reaches a constant value.

当在任何一个区域中发生负荷变化时，在互联系统中，仅仅在每一个发电机组的输出功率达到恒定值后，就会获得一个新的稳定运行点。

[13] For example, in the PJM power market in the eastern United States, the number of hours with active transmission line limit violations increased from 294 hours during the summer of 1998 to 548 hours during the summer of 1999.

例如，位于美国东部的 PJM 电力市场，传输线越限的总时长由 1998 年夏季的 294h 增加到 1999 年夏季的 548h。（PJM 是 Pennsylvania-New Jersey-Maryland，宾夕法尼亚-新泽西-马里兰的简称。）

[14] The objective of a transient stability study is to determine whether or not the machines will return to synchronous frequency with new steady-state power angles.

暂态稳定性研究的目标就是确定电机是否会转变到具有新的稳态功率角的同步频率。

[15] In today's large-scale power systems with many synchronous machines interconnected by complicated transmission networks, transient stability studies are best performed with a digital computer program.

现今，在带有许多通过复杂传输网络互联的同步机的大规模电力系统中，通过数字电脑将暂态稳定性的研究很好地呈现出来。

[16] However, where multiswings lasting several seconds are of concern, models of turbine-governors and excitation systems, as well as more detailed machine models can be employed to obtain accurate transient stability results over the longer time period.

然而，对于持续数秒的多次振荡，可以利用涡轮调速器和励磁系统的模型，以及更详细的机械模型来获取在较长一段时间内准确的暂态稳定结果。

Chapter 17 Generating Plants

In this chapter we give a simplified description of power sources, called generating plants (or generators). Most commonly the generating plants convert energy from fossil or nuclear fuels, or from falling water, into electrical energy.

17.1 Electric Energy

Electricity is only one of many forms of energy used in industry, homes, businesses, and transportation. It has many desirable features: it is clean (particularly at the point of use), convenient, relatively easy to transfer from point of source to point of use, and highly flexible in its use. [1] In some cases it is an irreplaceable source of energy.

Fig. 17-1 is a useful summary of electric energy sources and their transition to end users for the United States in 1996. The basic energy sources are shown on the left. The end users of the electricity are shown on the right. Only about one-third of the resource energy is converted into electricity; about two-thirds is lost as "waste heat". In some cases this heat is not wasted. It can be used for heating homes and offices or for some industrial processes.

In Fig. 17-1, the T & D losses are transmission and distribution losses (almost 10% of the net generation of electricity). Also, note the significant amount of nonutility energy generated in 1996. Changes in government energy policy have encouraged this growth. In the period from 1990 to 1995, nonutility power generation grew by 47%.

In 1996, most of the production has been in conventional steam plants. Conventional steam refers to steam generation by burning coal, petroleum, or gas. Approximately 3,000 billion kilowatt hours of electricity were produced. Of this, coal accounted for approximately 56%, petroleum 2%, natural gas 8% (totaling 66% for conventional steam), hydropower 11%, nuclear power 22%, and others, including gas turbines, about 1%. Note that nuclear and geothermal power plants also generate steam but not by burning fossil fuels.

In 1996, of the total installed generating capability of approximately 710 million kilowatts, some 63% was conventional (fossil fuel) steam, 14% was hydropower, 14% was nuclear, 8% was gas turbine, and others totaled about 1%. [2] Comparing these with the production figures given earlier, we see great differences in the utilization rates of the various sources. Nuclear power has the highest rate. Gas turbines and internal combustion engines are among the lowest in the rate of utilization. We will discuss the reasons in a moment.

First, it is interesting to calculate an overall utilization factor for 1996. Suppose that it had been possible to utilize the 710 million kilowatt capability full time. Then the plants would have produced $710 \times 10^9 \times 8760 = 6220 \times 10^{12}$ watthours in 1996. They actually produced 3078×10^{12}

watthours. Thus the annual capability factor or load factor was 3078/6220 = 0.49 or 49%. Why isn't the figure higher?

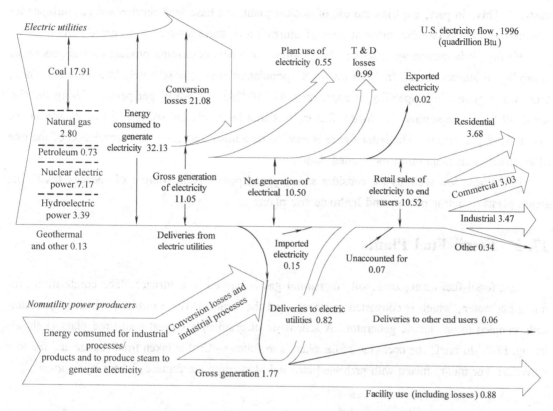

Fig. 17-1 U. S. Electricity Supply and Demand
(From Annual Energy Review 1996, U. S. Department of Energy.)

There are two main reasons. The first is that generating units are not always available for service. There is downtime because of maintenance and other scheduled outages; there are also forced outages because of equipment failures. The availability of fossil-fuel steam turbine units ranges from about 80% to about 92%.

The second reason involves a characteristic of the load. While there must be enough generating capability available to meet the requirements of the peak-load demand, the load is variable, with daily, weekly, and seasonal variations, and thus has a lower average value.[3] The daily variations are roughly cyclic with a minimum value (the baseload) typically less than one-half of the peak value. A typical daily load curve for a utility is shown in Fig. 17-2. The (weekly) capability factor for this particular utility is seen to be approximately 65%.

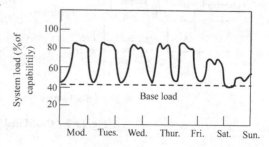

Fig. 17-2 Daily Load Output (typical week)

In meeting the varying load requirements, economic considerations make it desirable to utilize plants fully with low (incremental) fuel costs while avoiding the use of plants with high fuel costs. [4] This, in part, explains the use of nuclear plants for base-load service and gas turbines for peaking-power service; the different rates of utilization of these sources were noted earlier.

Finally, it is interesting to reduce the enormous numbers describing production and generating capability to human terms. In 1996 the U. S. population was approximately 265 million. Thus, there was a generating capability of approximately $710/265 = 2.68$kW per person. Using the figure 0.49 for the capability (or load) factor, this translates into an average use of energy at the rate 1.3kW per person. The latter figure is easy to remember and gives an appreciation of the rate of electricity consumption in the United States.

In the next few sections we consider some typical power plant sources of energy: fossil-fuel steam plants, nuclear plants, and hydroelectric plants.

17.2 Fossil-Fuel Plants

In a fossil-fuel plant, coal, oil, or natural gas is burned in a furnace. The combustion produces hot water, which is converted to steam, and the steam drives a turbine, which is mechanically coupled to an electric generator. A schematic diagram of a typical coal-fired plant is shown in Fig. 17-3. In brief, the operation of the plant is as follows: Coal is taken from storage and fed to a pulverizer (or mill), mixed with preheated air, and blown into the furnace, where it is burned. [5]

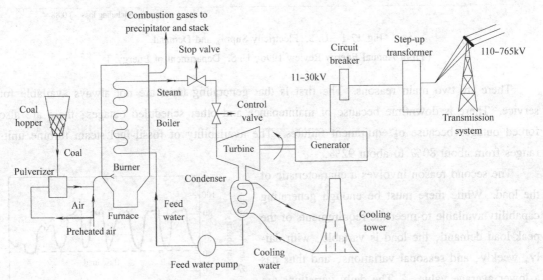

Fig. 17-3 Coal-Fired Power Station (schematic)

The furnace contains a complex of tubes and drums, called a boiler through which water is pumped; the temperature of the water rises in the process until the water evaporates into steam. The steam passes on to the turbine, while the combustion gases (flue gases) are passed through mechanical and electrostatic precipitators, which remove upward of 99% of the solid particles (ash) before being released to the chimney or stack.[6]

The unit just described, with pulverized coal, air, and water as an input and steam as a useful output, is variously called a steam-generating unit, or furnace, or boiler. When the combustion process is under consideration, the term *furnace* is usually used, while the term *boiler* is more frequently used when the water-steam cycle is under consideration. The steam, at a typical pressure of 3,500 psi and a temperature of 1,050°F, is supplied through control and stop (shutoff) valves to the steam turbine. The control valve permits the output of the turbine-generator unit (or turbogenerator) to be varied by adjusting steam flow. The stop valve has a protective function; it is normally fully open but can be "tripped" shut to prevent overspeed of the turbine-generator unit if the electrical output drops suddenly (due to circuit-break action) and the control valve does not close.

Fig. 17-3 suggests a single-stage turbine, but in practice a more complex multistage arrangement is used to achieve relatively high thermal efficiencies. A representative arrangement is shown in Fig. 17-4. Here, four turbines are mechanically coupled in tandem and the steam cycle is complex. In rough outline, high-pressure steam from the boiler (superheater) enters the high-pressure (HP) turbine. Upon leaving the HP turbine, the steam is returned to a section of the boiler (reheater) and then directed to the intermediate-pressure (IP) turbine. Leaving the IP turbine, the steam (at lower pressure and much expanded) is directed to the two low-pressure (LP) turbines. The exhaust steam from the LP turbines is cooled in a heat exchanger called a *condenser* and, as feedwater, is reheated (with steam extracted from the turbines) and pumped back to the boiler.

Finally, we get to the electric generator itself. The turbine turns the rotor of the electric generator in whose stator are embedded three (phase) windings. In the process mechanical power from the turbine drive is converted to three-phase alternating current at voltages in the range from 11kV to 30kV line to line at a frequency of 60Hz in the United States. The voltage is usually "stepped up" by transformers for efficient transmission to remote load centers.

Steam-driven turbine generators (a generator also called an *alternator* or *synchronous generator*) are usually two-pole or four-pole, turning at 3,600 r/min or 1,800 r/min, respectively, corresponding to 60Hz. The high speeds are needed to achieve high steam turbine efficiencies. At these rotation rates, high centrifugal forces limit rotor diameters to about 3.5ft for two-pole and 7ft for four-pole machines.

The average power ratings of the turbine-generator units we have been describing have been increasing, since the 1960s, from about 300MW to about 600MW, with maximum sizes up to about 1,300MW.[7] Increased ratings are accompanied by increased rotor and stator size, and with rotor diameters limited by centrifugal forces, the rotor lengths have been increasing. Thus, in the larger sizes, the rotor lengths may be five to six times the diameters. These slender rotors resonate at critical speeds below their rated speeds, and care is required in operation to avoid sustained operation at these speeds.[8]

Finally, we note some problems associated with the use of coal-fired power plants. Mining and transportation of coal, present safety hazards and other social costs. Coal-fired plants share environmental problems with some other types of fossil-fuel plants; these include acid rain and the greenhouse effect.

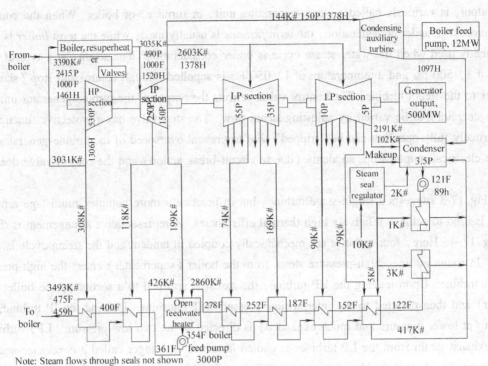

Fig. 17-4 Thermodynamic Cycle Diagram
(Adapted from McGraw-Hill Encyclopedia of Energy, 2nd ed, Sybil P. Parker. Ed, 1977.
Courtesy of McGraw-Hill Book Company, New York.)

17.3 Nuclear Power Plants

Controlled nuclear fission is the source of energy in a nuclear power plant. In the process of fission, heat is generated that is transferred to a coolant flowing through the reactor. Water is the most common coolant, but gases, organic compounds, liquid metals, and molten salts have also been used.

In the United States the two most common types of nuclear plants, collectively called *light-water reactors*, are the *boiling-water reactor* (BWR) and the pressurized-water *reactor* (PWR), and both use water as coolant.[9]

In the BWR the water is allowed to boil in the reactor core; the steam is then directed to the turbine. In the PWR there may be two or more cycles or loops linked by heat exchangers but otherwise isolated from each other. In all but the last stage (the secondary of which carries steam to the turbine) the water is pressurized to prevent steam generation.

Although it might appear that the only difference between a nuclear plant and a fossil-fuel plant is the way the steam is produced (i.e., by nuclear reactor/steam generator rather than furnace/boiler), there are some other differences. For example, nuclear steam generators are presently limited in their temperature output to about 600 °F (compared with about 1,000 °F for a fossil-fuel plant). This has a negative impact on thermal efficiency (30% instead of 40%) and on steam conditions in the turbines. There are, of course, major differences in the fuel cycle (supply and disposal) and in requirements for plant safety.

17.4 Hydroelectric Power Plants

Hydroelectric generation is an important source of power in the United States, accounting for approximately 14% of the installed generating capability and 11% of the energy production in 1996. Hydroplants are classified as *high head* (over 100ft) or *low head*. The term *head* refers to the difference of elevation between the upper reservoir above the turbine and the tail race or discharge point just below the turbine.[10] For very high heads (600 to 6,000ft) Pelton wheels or impulse turbines are used; these consist of a bucket wheel rotor with one or more nozzles directed at the periphery. For medium-high heads (120 to 1,600ft) Francis turbines are used. These turn on a vertical axis, with the water entering through a spiral casing and then inward through adjustable gates (i.e., valves) to a runner (i.e., turbine wheel) with fixed blades.[11] The low-head Kaplan-type turbine is similar but has adjustable blades in the runner. The efficiency of these turbines is fairly high, in the neighborhood of 88% to 93% when operating at their most efficient points.

We note a highly desirable feature of hydropower plants: the speed with which they may be started up, brought up to speed, connected to the power network, and loaded up. This process can be done in five minutes, in contrast to many hours in the case of thermal plants; the job is also much simpler and adaptable to remote control. Thus, hydropower is well suited for turning on and off at a dispatcher's command to meet changing power needs. When water is in short supply, it is desirable to use the limited available potential energy sparingly, for periods of short duration, to meet the peak-load demands. When water is plentiful, with the excess flowing over the spillway of the dam, base-loading use is indicated.

Hydropower can effectively meet peak demand in locations without suitable water flows by using *pumped storage*. Water is pumped from a lower reservoir to a higher one during off-peak times (generally at night) and the water is allowed to flow downhill in the conventional hydroelectric mode during times of peak demand. Off peak, the generators, operating as motors, drive the turbines in reverse in a pumping mode. The overall efficiency is only about 65% to 70%, but the economics are frequently favorable when one considers the economics of the overall system, including the thermal units. Consider the following. It is not practical to shut down the largest and most efficient thermal units at night, so they are kept online, supplying relatively small (light) loads. When operating in this mode the pumping power may be supplied at low incremental costs. On the other hand, at the time of peak demand the pumped storage scheme provides power that would otherwise have to be supplied by less efficient (older) plants. In a sense, from the point of view of the thermal part of the system, the pumped storage scheme "shaves the peaks" and "fills the troughs" of the daily load-demand curve. [12]

17.5 Wind Power Systems

Globally, wind power development is experiencing dramatic growth. According to the Global Wind Energy Council, GWEC, 15, 197MW wind turbine has been installed in 2006, an increase of 32% over 2005. The installation of the total global wind energy capacity is increased to 74, 223MW by the end of 2006 from 59, 091 MW of 2005. In terms of economic value, the wind energy sector has now become one of the important players in the energy markets, with the total value of new generating equipment installed in 2006 reaching US $23 billion or €18 billion. [13]

Europe continues to lead the world in total installed capacity. In 2006, the country having the highest total installed capacity is Germany with 20, 621MW. Spain and the United States are in second and third place, each with a little more than 11, 603MW installed. India is in fourth place, and Denmark ranks fifth. Asia experienced the strongest increase in installed capacity outside of Europe, with an addition of 3, 679MW, taking the total capacity over 10, 600MW, about half that of Germany. [14] The Chinese market was boosted by the country's new Renewable Energy Law. China has more than doubled its total installed capacity by installing 1, 347MW of wind energy in 2006, a 70% increase over 2005. This brings China up to 2, 604MW of capacity, making it the sixth largest market worldwide. It is expected that more than 1, 500MW will be installed in 2007. Growth in African and Middle Eastern market also picked up in 2006, with 172MW of new installed capacity—mainly in Egypt, Morocco, and Iran—bringing the total up to 441MW, a 63% growth.

The European Wind Energy Association (EWEA) has set a target to satisfy 23% European electricity needs with wind by 2030. The exponential growth of the wind industry reflects the increasing demand for clean, safe and domestic energy and can be attributed to government policies associated with the environmental concerns, research and development of innovative cost-reducing technologies.

The large scale development of wind power results in the wind turbines/farms becoming a significant part of the generation capacity in some area, which requires that the power system treats the wind turbines/farms like a power source, not only an energy source. The wind power penetration would result in variations of load flows in the interconnected systems, as well as re-dispatch of conventional power plants, which may causes the reduced reserve power capacity. Some actions become necessary to accommodate large scale wind power penetration. For example, the electric grid may need an expansion for bulk electricity transmission from offshore wind farms to load centers, and it may require reinforcement of existing power lines or construction of new power lines, installation of Flexible AC Transmission system (FACTs) devices, etc.

The electrical power produced by wind turbine generators has been increasing steadily, which directly pushes the wind technology into a more competitive area. Basically a wind turbine consists of a turbine tower, which carries the nacelle, and the turbine rotor, consisting of rotor blades and hub. Most modern wind turbines have three rotor blades usually placed upwind of the tower and the nacelle. On the outside the nacelle is usually equipped with anemometers and a wind wane to measure the wind speed and direction, as well as with aviation lights.[15] The nacelle contains the key components of the wind turbine, e.g. the gearbox, mechanical brakes, electrical generator, control systems, etc. The wind turbines are not only installed dispersedly on land, but also combined as farms with capacities of hundreds MWs, which are comparable with modern power generator units. Consequently, their performance could significantly affect power system operation and control. The main components of a modern wind turbine system are illustrated in Fig. 17-5, including the turbine rotor, gear box, generator, transformer and possible power electronics.

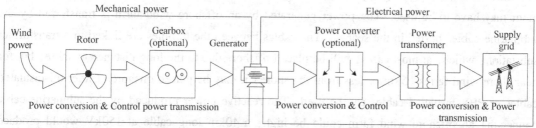

Fig. 17-5 Main Components of a Wind Turbine System

The conversion of wind power to mechanical power is done aerodynamically. The available power depends on the wind speed but it is important to be able to control and limit the power at higher wind speed to avoid damage. The power limitation may be done by stall control (the blade position is fixed but stall of the wind appears along the blade at higher wind speed), or active stall (the blade angle is adjusted in order to create stall along the blades) or pitch control (the blades are turned out of the wind at higher wind speed), which result in power curves as shown in Fig. 17-6.

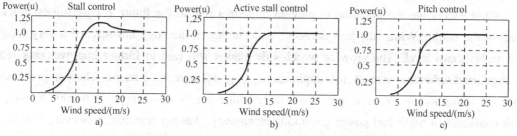

Fig. 17-6 Power Characteristic of Fixed Speed Wind Turbines
a) Stall Control   b) Active Stall Control   c) Pitch Control

Mainly three types of typical wind generator systems exist. The first type is a constant-speed wind turbine system with a standard squirrel-cage induction generator (SCIG) directly connected to the grid. The second type is a variable speed wind turbine system with a doubly fed induction generator (DFIG). The power electronic converter feeding the rotor winding has a power rating of approximately 30% of the rated power; the stator winding of the DFIG is directly connected to the grid. The third type is a variable speed wind turbine with full-rated power electronic conversion system and a synchronous generator or a SCIG. A multi-stage gearbox is usually used with the first two types of generators. Synchronous generators, including permanent magnet generator, may be direct driven, though a low ratio gear box system, one or two stage gearbox, becomes an interesting option.

The suitable voltage level is related to the amount of power generated. A modern wind turbine is often equipped with a transformer stepping up from the generator terminal voltage, usually a voltage below 1kV, to a medium voltage around 20kV or 30kV, for the local electrical connection within a wind farm. If the wind farm is large and the distance to the grid is long, a transformer may be used to further step up the medium voltage in the wind farm to a high voltage at transmission level. For example, for large onshore wind farms at hundreds of MW level, high voltage overhead lines above 100kV are normally used. For offshore wind farms with a long distance transmission to an onshore grid, a high voltage submarine cable with a lead sheath and steel armour may have to be used. The power generated by an offshore wind farm is transferred by the submarine cables buried in the seabed. The cables between the turbines are linked to a transformer substation, which, at most cases, will be placed offshore due to the long distance to shore, but for near shore wind farms (5km or less from the shore) it may be placed onshore. Either oil-insulated cables or PEX-insulated cables can be used. The reactive power produced by the submarine cable connecting an offshore wind farm could be high, a 40km long cable at 150kV would produce around 100Mvar, and reactors may be needed to compensate the reactive power produced by the cable. For long distance transmission, the transmission capacity of the cables may be mainly occupied by the produced reactive power. In this situation high voltage direct current (HVDC) transmission techniques may be used. The new technology, voltage source converter based HVDC system, provides new possibilities for performing voltage regulation and improving dynamic stability of the wind farm as it is possible to control the reactive power of the wind farm and perhaps keep the voltage during a fault in the connected transmission systems.

Wind energy has the potential to play an important role in the future energy supply in many areas of the world. Within the past decades, wind turbine technology has reached a very reliable and sophisticated level. The growing world-wide market will lead to further improvements, such as large wind turbines or new system applications, e.g. offshore wind farms. These improvements will lead to further cost reductions and over the medium term wind energy will be able to compete with conventional fossil fuel power generation technology. Further research, however, will be required in many areas, and the skills will grow more and more mature.

17.6 Photovoltaic Systems

A photovoltaic system is an integrated assembly of modules and other components, designed to convert solar energy into electricity to provide a particular service, either alone or in combination with a back up supply. [16] A module is the basic building block of a photovoltaic generator. It is defined as the smallest complete, environmentally protected assembly of series connected solar cells. The cells are encapsulated between a transparent window and a moisture proof backing to insulate them electrically, as well as from the weather and accidental damage. Digital leads are provided for connecting it to other modules or components or the load. The modules in a PV array are connected in series strings to provide the required voltage, and if one string is not enough to provide the required power, two or more strings are connected in parallel.

In its simplest form, a standalone photovoltaic system consists of an array of one or more photovoltaic modules supplying the load directly (Fig. 17-7). Such a system can be used for water pumping, battery charging, etc. The addition of an inverter makes the system suitable for domestic supplies when the load consists of AC appliances (Fig. 17-8). Moreover, to carry the load during the night or during periods of low irradiance, a storage battery with a charge regulator must be added to the basic system (Fig. 17-9 and Fig. 17-10).

So, we have seen above that, in a complete photovoltaic system, the solar cell, inverter, regulator, battery, rectifier, etc. are the important components of the system.

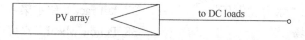

Fig. 17-7 Standalone DC System without Battery

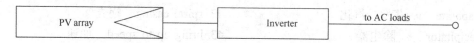

Fig. 17-8 Standalone AC System without Battery

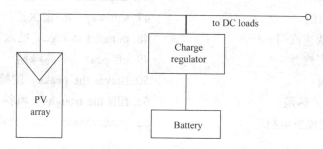

Fig. 17-9 Standalone DC System with Battery

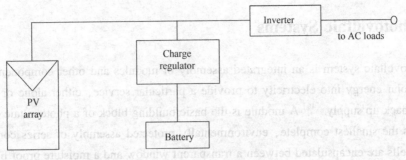

Fig. 17-10 Standalone AC System with Battery

New Words and Expressions

1. hydropower   n. 水电，水能
2. geothermal   adj. 地热的
3. installed generating capability   装机容量
4. internal combustion engine   内燃机
5. downtime   n. （由于检修）停机时间
6. peak-load   高峰负荷
7. base load   基本负荷
8. daily load curve   日负荷曲线
9. utility   n. 公共事业电力企业
10. incremental   adj. 逐渐增加的
11. hydroelectric plant   水力发电厂
12. storage   n. 贮煤场
13. pulverizer   n. 粉碎机
14. mill   n. 磨粉机
15. drum   n. 汽包，炉筒
16. evaporate   v. 蒸发，汽化
17. precipitator   n. 除尘器
18. stack   n. 通风管
19. psi   abbr. 磅/平方英寸
20. coal hopper   煤斗
21. stop valve   主汽门
22. control valve   调速汽门
23. condenser   n. 凝汽器
24. steam seal   汽封
25. enthalpy   n. 焓，热函
26. alternator   n. 交流发电机
27. centrifugal force   离心力
28. resonate   v. 谐振
29. greenhouse effect   温室效应
30. nuclear fission   核裂变
31. coolant   n. 冷却剂
32. reactor   n. 反应堆
33. organic compound   有机化合物
34. molten salt   熔盐
35. elevation   n. 高度，海拔
36. reservoir   n. 水库
37. tail race   尾水渠
38. Pelton wheel   水斗式水轮机
39. impulse turbine   冲击式水轮机
40. bucket wheel   斗式链轮，斗轮
41. Francis turbine   轴向辐流式水轮机
42. spiral casing   蜗壳
43. bring up to speed   加速
44. load up   带起负荷
45. remote control   遥控
46. dispatcher   n. 调度员
47. spillway   n. 溢洪道
48. pumped storage   抽水蓄能
49. off-peak   非高峰期
50. shaves the peaks   削峰
51. fills the troughs   填谷
52. photovoltaic   adj. 光电（池）的

Notes

[1] It has many desirable features: it is clean (particularly at the point of use), convenient, relatively easy to transfer from point of source to point of use, and highly flexible in its use.

它有许多可取的特点：它是清洁的（特别是在使用处）、方便的，从源端到使用端的转移也是相对方便的，使用起来也是高度灵活的。

[2] In 1996, of the total installed generating capability of approximately 710 million kilowatts, some 63% was conventional (fossil fuel) steam, 14% was hydropower, 14% was nuclear, 8% was gas turbine, and others totaled about 1%.

在1996年，总发电装机容量约为7.1亿kW，其中63%为传统发电（蒸汽机），14%为水电，14%为核电，8%为燃气轮机发电，其他占1%。

[3] While there must be enough generating capability available to meet the requirements of the peak-load demand, the load is variable, with daily, weekly, and seasonal variations, and thus has a lower average value.

负荷随着日、周、季节的变化而变化，因此具有较低的平均值，然而系统必须有足够的发电量来满足高峰负荷的需求。

[4] In meeting the varying load requirements, economic considerations make it desirable to utilize plants fully with low (incremental) fuel costs while avoiding the use of plants with high fuel costs.

在满足负荷变化的需求时，应考虑经济效益，理想的是充分利用低（微增率）燃料费用的电厂，同时避免使用高燃料费用的电厂。

[5] In brief, the operation of the plant is as follows: Coal is taken from storage and fed to a pulverizer (or mill), mixed with preheated air, and blown into the furnace, where it is burned.

简言之，工厂的操作如下：从仓库中取出煤，并且放入磨煤机中，与预热的空气混合，并且吸入炉内，在炉内燃烧。

[6] The steam passes on to the turbine, while the combustion gases (flue gases) are passed through mechanical and electrostatic precipitators, which remove upward of 99% of the solid particles (ash) before being released to the chimney or stack.

蒸汽通过涡轮，而燃烧气体（烟气）通过机械和静电除尘，在被释放到烟囱前可以去除99%以上的固体颗粒。

[7] The average power ratings of the turbine-generator units we have been describing have been increasing, since the 1960s, from about 300MW to about 600MW, with maximum sizes up to about 1,300MW.

自20世纪60年代以来，我们一直在描述的汽轮发电机组平均额定功率一直在增加，从300MW到600MW，最大约为1300MW。

[8] These slender rotors resonate at critical speeds below their rated speeds, and care is required in operation to avoid sustained operation at these speeds.

这些细长的转子在低于其额定转速的某个临界速度下会发生谐振，因此应避免持续运行在这种转速下。

[9] In the United States the two most common types of nuclear plants, collectively called *light-water reactors*, are the boiling-water reactor (BWR) and the *pressurized-water reactor* (PWR), and both use water as coolant.

在美国，被统称为轻水反应堆的核电站通常有两种主要类型，分别是是沸水反应堆（BWR）和压水堆（PWR），并且都以水作为冷却液。

[10] The term *head* refers to the difference of elevation between the upper reservoir above the turbine and the tail race or discharge point just below the turbine.

水头是指处于水轮机上方较高处的水库与紧靠水轮机排水口处水尾之间的水位差。

[11] These turn on a vertical axis, with the water entering through a spiral casing and then inward through adjustable gates (i.e., valves) to a runner (i.e., turbine wheel) with fixed blades.

通过蜗壳进入的水流向内通过调节阀门冲击到装有固定叶片的转子（即水轮机轮盘）上，从而驱动竖直安放的轴。

[12] In a sense, from the point of view of the thermal part of the system, the pumped storage scheme "shaves the peaks" and "fills the troughs" of the daily load-demand curve.

在某种意义上，从系统的火电角度来说，抽水储存的运行模式对日负荷曲线起到了一个削峰填谷的作用。

[13] In terms of economic value, the wind energy sector has now become one of the important players in the energy markets, with the total value of new generating equipment installed in 2006 reaching US＄23 billion or 18 billion.

就经济价值而言，风能现在已经成为能源领域的重要成分之一。在 2006 年，新安装的发电设备就达 230 亿美元（180 亿欧元）。

[14] Asia experienced the strongest increase in installed capacity outside of Europe, with an addition of 3,679MW, taking the total capacity over 10,600MW, about half that of Germany.

除欧洲外，亚洲的装机容量增长最快，新增 3679MW，总装机容量已达 10 600MW，约为德国装机容量的一半。

[15] On the outside the nacelle is usually equipped with anemometers and a wind wane to measure the wind speed and direction, as well as with aviation lights.

在机舱的外面通常会配备风速计和一个风向标，以及航空灯来测量风的速度和方向。

[16] A photovoltaic system is an integrated assembly of modules and other components, designed to convert solar energy into electricity to provide a particular service, either alone or in combination with a back up supply.

光伏系统是由模块和其他元件集成的，它将太阳能转化为电能，可以单独或与备用电源一起提供特殊的供电服务。

Chapter 18　Concepts and Models for Microgeneration and Microgrids

18.1　Introduction

In the last years, the electric power system experienced a massive adoption of Distributed Generation (DG) resources. [1] The change of paradigm was more perceptible due to the connection of large amounts of DG sources to Medium Voltage (MV) distribution networks. However, recent technological developments are contributing to the maturation of some DG technologies, such that they are becoming suitable to be connected to Low Voltage (LV) distribution grids. Regarding a scenario characterized by a massive integration of DG in the network, several technical issues need to be tackled. In order to face these challenges and to realize the potential benefits of DG resources, it is imperative to develop a coordinated strategy for its operation and control, together with electrical loads and storage devices. [2] A possible approach is the development of the MicroGrid (MG) concept. Within the scope of this dissertation, a MG comprise a LV distributed system with small modular generation technologies, storage devices and controllable loads, being operated connected to the main power network or islanded, in a controlled coordinated way. The small modular generation technologies, also denominated by MicroSources (MS), are small units with electrical power ratings of less than 100k.We, most of them with power electronic interfaces, and exploiting either Renewable Energy Sources (RES) or fossil fuels in high efficiency local Combined Heat and Power (CHP) applications. The types of MS suitable to be used in a MG are: micro gas turbines, fuel cells (different types) photovoltaic panels, small and micro wind generators, together with storage devices such as batteries, flywheels or supercapacitors.

The MG concept is a natural evolution of simple distribution networks with high amounts of DG, since it offers considerable advantages for network operation due to its additional control capabilities. [3] The formation of active LV networks through the exploitation of the MG concept can potentially provide a number of benefits to the Distribution Network Operator (DNO) and to the end-user. These benefits are in-line with the major factors contributing to a massive adoption and deployment of DG technologies:

1) MG operation is based in a large extent on RES and MS characterised by zero or very low emissions. In addition, the reduction of distribution system losses resulting from DG integration contributes also for decreasing Greenhouse Gas (GHG) emissions.

2) MG will exploit either RES or fossil fuels in high efficiency local micro-CHP applications. The intensive use of micro-CHP applications contributes to the raise of overall energy systems efficiency to levels far beyond what is possible to achieve with central power stations. On the

other hand, the exploitation of local RES will contribute to the reduction of the dependence on imported fossil fuel and to increase energy security.

3) A consumer integrated in a MG will be able to act both as a buyer or seller of thermal and electrical energy. This flexibility potentially allows the development of a generation system with a higher overall efficiency and being able to more responsive to customer needs.

4) A well designed MG is capable of enhancing power system reliability at the customer level, since two independent sources (the MV distribution grid and DG units) can be used to supply the electrical loads.[4] Also, the transmission system dynamic stability can be improved under a scenario of provision of ancillary services by DG. Additional benefits such as voltage support or enhanced power quality can also be provided by MG.

5) The installation of power generation closer to the loads lowers overall system transmission and distribution losses, and can be an interesting solution in order to face the growing power needs. In fact, DG can potentially prevent or defer the investments required for upgrading or building additional central power generation units or transmission and distribution infrastructures.

Achieving a coordinated control of the MG cell in order to provide the required flexibility of operation is a challenging task and can be realized only by means of the hierarchical control structure to be developed according to MG specific requirements. The exploitation of a centralized control strategy would require multiple high data rate bidirectional communication infrastructures, powerful central computing facilities and a set of coordinated control centers. Therefore, such an approach is not attractive for a practical realization due to the inherent high costs of a high data rate and extremely reliable telecommunication and control infrastructure. Additionally, and in order to guarantee an efficient and reliable control of the MG under abnormal operating conditions, the most practical approach should rely on a network of local controllers in order to handle the resulting transient phenomena and guarantee MG survival.[5] Therefor, MG control strategies should be based on a network of controllers with local intelligence. The information to be exchanged should be limited to the minimum necessary to achieve the optimization of MG operation when the MG is connected to the main grid, to assure stable operation during islanding conditions and to implement local recovery functionalities after a general failure (local Black Start functionalities).

Regarding MG operation and control specificities, they are inherently associated with a set of issues, including safely, reliability, voltage profile, power quality, protection, unbalance/asymmetry and non-autonomous/autonomous operation. In particular, the operation and control issues of a MG are challenging problems due to the very specific nature of the new electrical power system under consideration. Therefore, adequate modeling of MS, storage devices and power electronic interfaces is also addressed in this chapter, since it is the first step required in order to evaluate the dynamic interactions that may occur within a MG and to evaluate the feasibility of the control strategies to be derived.[6]

This chapter contains therefore the presentation of the MG concept and its overall control and management architecture, together with the description of the models adopted to MS control structures and their corresponding power electronic interfaces are also presented. Such modeling is of paramount importance for the success of the development of the research work of this dissertation.

18.2 The Foundation of Microgrid Concept

Europe, North America and Japan are leading the revolution being faced in the conventional electric power systems operation paradigm, by actively promoting research, development and deployment of MG. [7] The MG concept was originally introduced in the United States by the Department of Energy, who have actively supported considerable work in the area more specifically, the Consortium for Electric Reliability Technology Solutions (CERTS) was founded in 1999 to research, develop and disseminate new methods, tools and technologies in order to protect and enhance the reliability of the United States electric power system and the efficiency of competitive electricity markets. The CERTS electricity reliability research covers several areas, one of which is devoted to Distributed Energy Resources (DER) integration, by developing tools and techniques to maintain and enhance the reliability of electricity service through a cost-effective, decentralized electricity system based on high penetrations of DER.

Since the early beginnings, a "fit and forget" policy was followed by distribution companies when connecting DG to the system. However, a true integration policy is needed, through which a system perspective is used to capture the potential benefits that may arise to customers and to the utilities form increasing DG penetration levels. The CERTS MG concept is an advanced approach aiming to the effective large scale DG integration in distribution systems. The conventional approach for DER integration is focused on the impact resulting from the connection of a small number of DG to the grid, as it can be seen by analysing the IEEE P1574 standard (IEEE standard for Interconnecting Distributed Resources with Electric Power Systems). The focus of this standard is to assure DG is quickly disconnected following the event of grid disturbances. On contrary, the MG is regarded as an aggregation of loads; MS and storage devices that can be operated connected to or separate from the main electricity grid. One of the key functionalities of the MG is to seamlessly separate from a normal utility service is restored. Following system disturbances, DG and electrical loads can automatically separate from the main power system, therefore isolating the MG from the disturbance affecting the system. The adoption of intentional islanding practices can contribute to increase local reliability indexes to levels higher than the ones ensured by the electric power system as a whole.

The CERTS MG concept assumes an aggregation of loads and MS operating in a single system and providing both power and heat to local consumers. The majority of MS should be power electronic interfaced in order to provide the required flexibility, assure MG operation as a single aggregated system and to achieve a plug-and-play simplicity for each MS. From the bulk power system perspective, the MG can establish contractual agreements for energy and possibly other services in a similar way to what happens with ordinary consumers or power producers. [8] From the technical point of view, MG connection to the distribution grid should satisfy at least minimum requirements that other conventional equipments are obliged to satisfy. However, the CERTS emphasizes MG flexibility should be explored further and should not be limited to not jeopardizing the surrounding electrical power system by behaving as a "good citizen".

The CERTS MG exploits mainly DG technologies with power electronic interfaces in order to provide the required flexibility and controllability. In principle, no limitations are imposed to DG ratings; nevertheless, pragmatic reasons related to availability and controllability, lead to focusing the effort on LV grids to which microturbines with power ratings less than 500 kW are connected. Although it is not the case, other emerging MS technologies like fuel cells can also be considered as candidates to be integrated in the MG. The possibility of extending the MG concept to large systems is also referred. In case of large sites, it is suggested to divide the loads in many controllable units (buildings, industrial sites, etc.) and exploit the distribution system for the interconnection of several MG in order to supply the entire system.[9]

The basic CERTS architecture is shown in Fig. 18-1, where it is represented a typical LV distribution system with several radial feeders (in these case A, B and C). The LV side of the distribution transformer is the Point of Common Coupling (PCC) between the MG and the distribution

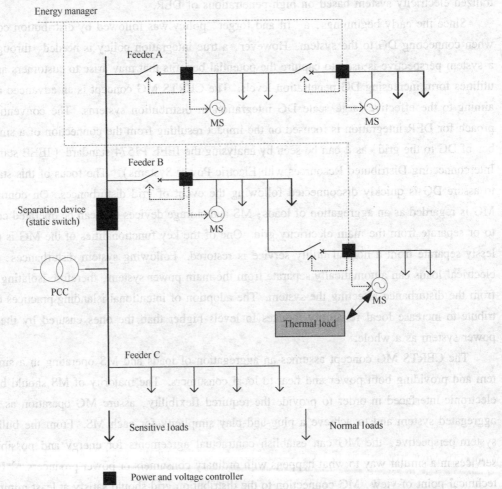

Fig. 18-1 The CERTS MG Architecture

Chapter 18　Concepts and Models for Microgeneration and Microgrids

system and it is used to define the boundary between both systems.[10] At the PCC, the MG should comply with existing interface requirements (for example, the IEEE P1547 standard). As it shown in Fig. 18-1, the key elements of the CERTS MG architecture are:

(1) Microsource Controller (Power and Voltage Controller). The basic MG operation relies on the MS controller in order to perform the following actions: control feeder power flow according to pre-defined criteria, voltage control at the MS connection point, load sharing among MS following MG islanding and MG synchronization with the upstream MV network. The response time of the MS controller is very fast (in the order of milliseconds) and the control functions are performed using only measurements available locally at the controller connection point. Running in a larger time frame, the Energy Manager, a kind of central control system, defines operational strategies to achieve an optimal management of the entire MG. This requires a communication channel to be established between the Energy Manager and the MS controllers. The MG control architecture confers the plug-and-play characteristic to MS, that is: MS can be connected to the MG without requiring modification in the control and protection functions of the units already making part of the system.

(2) Energy Manager. The Energy Manager is responsible for managing the MG operational control, by periodically providing the adequate power and voltage dispatches to the MS controllers according to pre-defined criteria such as: reduction of MG losses, maximizing MS operation efficiency, satisfying the contractual agreements at the PCC, etc. The control functions in the Energy Manager are executed with a periodicity of a few minutes.

(3) Protections. The MG protection scheme should ensure an adequate response to faults occurring in the upstream MV network or in the MG itself in order to provide the required reliability levels to the critical loads or MG sections. The isolation speed is dependent on the specific customer loads on the MG (in some cases, voltage sag compensation can be used without separation from the distribution system in order to protect the critical loads). If the fault is located within the MG, the protection coordinator should isolate the smallest possible section of the radial feeder in order to eliminate the fault. The development of the protection system requires special attention due to the massive presence of power electronic converters, which have reduced capability to provide large fault currents.[11] Eventually, it will be necessary to develop alternative methods to the conventional over-current protection schemes used in distribution systems.

In addition to MG islanding, the CERTS MG architecture allows the following operation modes when the MG is interconnected with the upstream system:

(4) Unit Power Control Configuration. Each MS control its own power injection and the voltage magnitude at the connection point. This operation mode is specially envisaged for MS associated to thermal loads, since electric power production is driven by thermal loads requirements.

(5) Feeder Flow Control Configuration. The MS are operated in order to control the voltage magnitude at the point of connection and to maintain a schedule power flow in strategic points of the feeder. In this case, load variations in the feeder are picked up by the MS.

(6) Mixed Control Configuration. In this case some MS regulate their output power, while

others control feeder flows.

18.3 The Microgrid Operational and Control Architecture

In the European Union, the first major effort devoted to MG was initiated with the Fifth Framework Program (1998-2002), which funded the R&D project entitled "MICROGRIDS—Large Scale Integration of Micri-Generation to Low Voltage Grids". Within this project, a MG can be defined as a LV network (e. g., a small urban area, a shopping centre or an industrial park) plus its loads and several small modular generation systems connected to it, providing both power and heat to local loads. A MG may also include storage devices (such as batteries, flywheels or supercapacitors) and network control and mamagement systems. The MG concept developed within the MICROGRIDS project, and followed in this dissertation, is shown in Fig. 18-2. The figure illustrates a typical LV distribution network connected to the secondary winding of a MV/LV distribution transformer. This MG example includes:

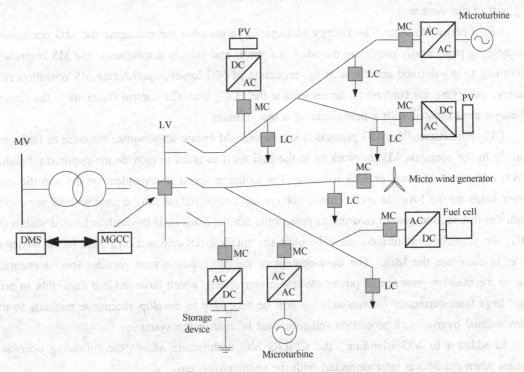

Fig. 18-2 MG Architecture, Comprising MS, Storage Devices and a Hierarchical Control and Management System

1) Several feeders supplying electrical loads.
2) Microgeneration systems based on renewable energy sources such as Photovoltaic (PV) or micro wind generators and fuel-based MS in CHP applications (a microturbine and a fuel cell).
3) Storage devices.
4) A hierarchical-type management and control scheme supported by a communication infrastructure, in order to ensure all the elements of MG are aggregated in a single cell that is interfaced

to the electrical power system in a similar way as ordinary consumers or DG sources.

A MG cell is intended to operate connected to or isolated from the upstream MV network, allowing the definition of the following operation modes:

(1) Normal Interconnected Mode: the MG is connected to the upstream MV network, either being totally or partially supplied by it (depending on the dispatching procedures used to operated the MS) or injecting some amount of power into the main system (in case the relation between the MS production level and the total MG consumption allows this type of operation).

(2) Emergency Mode: following a failure in the upstream MV network, or due to some planned actions (for example, in order to perform maintenance actions) the MG can have the ability to smoothly move to islanded operation or to locally exploit a service restoration procedure in the advent of a general blackout. In both cases, the MG operates autonomously, in a similar way to the electric power systems of the physical islands.

In order to achieve the desired flexibility, the MG system in centrally controlled and managed by the MicroGrid Central Controller (MGCC), installed in the LV side of the hierarchical level. [12] The second hierarchical control level comprises MS and storage devices being locally controlled by a Microsource Controller (MC) and the electrical loads or group of loads being controlled by a Load Controller (LC). The proper operation and control of the entire system requires communication and interaction between the referred hierarchical control levels as follows:

1) The LC and MC, on one hand, as interfaces to control loads through the application of the interruptability concept, and MS active and reactive power production levels.

2) The MGCC, on the other hand, as the central controller responsible for an adequate technical and economical management of the MG according to pre-defined criteria, by providing setpoints to MC and LC.

It is also expected the MGCC to be able to communicate with the Distribution Management System (DMS), located upstream in the distribution network, contributing to improve the management and operation of the MV distribution system through contractual agreements that can be established between the MG and the DNO. In order to enable this scenario, the conventional approaches to DMS need to be enhanced with new features related to MG connected on the feeders. The issues of autonomous and non-autonomous operation of the MG and the related exchange of information are examples of new important issues to be tackled in the near future. [13]

The MC can be housed within the power electronic interface of the MS. It responds in milliseconds and uses local information and the demands from the MGCC to control the MS during all events. The MC will have autonomy to perform local optimization of the MS active and reactive power production, when connected to the power grid, and fast load-tracking following an islanding situation. LC also need to be installed at the controllable loads to provide load control capabilities following demands from the MGCC, under a Demand Side Management (DSM) policy, or in order to implement load shedding functionalities during emergency situations. By exploiting the proposed architecture, the required MG operation and control functionalities that assure a stable operation in the first moments subsequent to transients are implemented based only on information available locally at the MC and LC terminals. Operational strategies intended for global MG optimization will run periodically (few minutes) in the MGCC and the resulting dispatch (voltage set-points, active and reactive power set-points, loads to be shed or deferred in time, etc.) will be communicated to local controllers (MC and LC) in a second stage corresponding to a larger time frame.

The MGCC heads the technical management of the MG. During the Normal Interconnected Mode, the MGCC collects information from the MC and LC in order to perform a number of functionalities. A key functionality to be installed in the MGCC is forecasting of local loads and generation.[14] The MGCC will be responsible for providing system load forecasts (electric and possibly heat). It will also forecast in a simpler manner power production capabilities (exploiting information coming from wind speed, insulation levels, etc.) and it will use electricity and gas costs information and grid needs, together with security concerns and DSM requests to determine the amount of power that the MG should absorb from the distribution system, optimizing the local production capabilities. The defined optimized operating scenario is achieved by controlling the MS and controllable loads in the MG in terms of sending control signals to the field.

In the Emergency Mode, an immediate change in the output power control of the MS is required, as they change from a dispatched power mode to one controlling frequency and voltage of the islanded section of the network.[15] Under this operating scenario, the MGCC performs an equivalent action to the secondary control loops existing in the conventional power systems: after the initial reaction of the MC and LC, which should ensure MG survival following islanding, the MGCC performs the technical and economical optimization of the islanded system. It is also important to the MGCC to have accurate knowledge of the type of loads in the MG in order to adopt the most convenient interruption strategies under emergency conditions. Being an autonomous entity, the MG can also perform local Black Start (BS) functions under certain conditions. If a system disturbance provokes a general blackout such that the MG was not able to separate and continue in islanding mode, and if the MV system is unable to restore operation in a specified time, a first step in system recovery will be a local BS. The strategy to be followed will involve the MGCC, the MC and the LC using predefined rules to be embedded in the MGCC software.

In the proposed MG architecture, some communication capabilities need to established between the MGCC and the local controllers, namely for control and operational optimization purposes. The amount of data to be exchanged between network controllers is small, since it includes mainly messages containing set-points to LC and MC, information requests sent by the MGCC to LC and MC about active and reactive powers and voltage levels and messages to control MG switches. Also, the short geographical span of the MG may aid establishing a communication infrastructure using low cost communications. The adoption of standard protocols and open technologies allows designing and developing modular solutions using off-the-shelf, low cost, widely available and fully supported hardware and software components. These solutions provide flexibility and scalability for future low cost implementations.

Having in mind the need to reduce costs in telecommunication infrastructures, an interesting solution could be the exploitation of power lines communication purposes (using the Power Line Communication (PLC) technology). In this case, the connectivity characteristics of the power grid provide the appropriate physical link between the different elements of the MG control system. Therefore, a careful analysis and evaluation of the power grid as the physical path for the communication system was performed within the MICROGRIDS project, namely in what concerns the characteristics of the physical transmission channels. The attenuation of the communication signal, as it propagates along the cable, can be too high if the communications path is too long. When evaluating the quality of the communications channel, another important factor must also be considered: the level and nature of the interfering signals that are present at the input of the receiver.[16] A number of interfering signals are generated by the connected loads and, hence, have different origins and characteristics, e.g. periodic signals (related to and/or synchronous with the power frequency), impulse-type signals and noise-like signals. If the amount of interfering signals is too large, with respect to signal distortion, then the receiver will have difficulties to reproduce the original information with sufficient accuracy.

Concerning the communication protocols to be used, a TCP/IP based transport protocol will provide extra functionality, flexibility and scalability, specially in terms future system evolutions like the exploitation of more complex scenarios (for example, multiple MG) or the use of the physical communication layer of the MG in order to support also other communication services. Additionally, the choice of a TCP/IP as the transport protocol makes possible the choice of any physical infrastructure. Therefore, it was proposed to support the MG control architecture without any specific dependence on the access technology, since a number of alternatives support the basic MG communication requirements.

New Words and Expressions

1. Distributed Generation(DG)
   分布式发电
2. Medium Voltage (MV) 中压
3. Low Voltage (LV) 低压
4. MicroSource (MS) 微电源
5. Renewable Energy Sources (RES)
   可再生能源
6. Combined Heat and Power (CHP)
   热电联产
7. Distribution Network Operator (DNO)
   配电网运营商
8. paradigm *n.* 范例
9. scenario *n.* 方案
10. efficiency *n.* 效率
11. exploitation *n.* 开发
12. installation *n.* 安装
13. coordinated control 协调控制
14. infrastructure *n.* 基础设施
15. intelligence *n.* 理解力
16. functionality *n.* 功能
17. autonomous *adj.* 自治的
18. storage devices 存储设备
19. power electronic interfaces
    电力电子接口
20. Consortium for Electric Reliability Technology Solutions (CERTS)
    电力可靠性技术解决联盟
21. Distributed Energy Resources (DER)
    分布式能源
22. conventional *adj.* 传统的

23. aggregation *n.* 集成
24. functionality *n.* 功能
25. disturbance *n.* 干扰
26. microturbine *n.* 微型气轮机
27. fuel cells 燃料电池
28. distribution system 配电系统
29. power flow 潮流
30. synchronization *n.* 同步
31. optimal management 最优管理
32. plug-and-play *adj.* 即插即用的
33. periodicity *n.* 周期性
34. critical load 临界负荷
35. power electronic converters 电力电子变换器
36. interconnect *v.* 互连
37. upstream system 上行系统
38. thermal load 热负荷
39. voltage magnitude 电压幅值
40. generation system 发电系统
41. infrastructure *n.* 基础设施
42. blackout *n.* 停电
43. centrally control 集中控制
44. hierarchical control 分层控制
45. interruptability 可中断性
46. operating scenario 操作方案
47. scalability *n.* 可扩展性
48. interfering signal 干扰信号
49. impulse-type signal 脉冲信号
50. transport protocol 传输协议

Notes

[1] In the last years, the electric power system experienced a massive adoption of Distributed Generation (DG) resources.

在过去的这些年中,电力系统大规模地采用了分布式发电资源。

[2] In order to face these challenges and to realize the potential benefits of DG resources, it is imperative to develop a coordinated strategy for its operation and control, together with electrical loads and storage devices.

为了面对这些挑战和实现分布式发电资源的潜在优势,制定一个计及电力负荷和储能设备的协调运行和控制的策略势在必行。

[3] The MG concept is a natural evolution of simple distribution networks with high amounts of DG, since it offers considerable advantages for network operation due to its additional control capabilities.

微电网的概念是一个包含大量分布式发电的简单分布网络的自然进化,由于其额外的控制能力,它为电网运行提供了相当大的优势。

[4] A well designed MG is capable of enhancing power system reliability at the customer level, since two independent sources (the MV distribution grid and DG units) can be used to supply the electrical loads.

在客户层面上,一个设计得比较好的微电网能够提高电力系统的可靠性,因为两个独立的电源(中压配电网和分布式发电网络)可以用来提供电力负荷。

[5] Additionally, and in order to guarantee an efficient and reliable control of the MG under abnormal operating conditions, the most practical approach should rely on a network of local controllers in order to handle the resulting transient phenomena and guarantee MG survival.

此外,在异常操作条件下,为了保证微电网高效可靠地控制,同时也为了处理由此而导致的暂态现象和保证微电网可靠运行,最实用的方法应该是依靠本地控制器的网络。

[6] Therefore, adequate modeling of MS, storage devices and power electronic interfaces is also addressed in this chapter, since it is the first step required in order to evaluate the dynamic interactions that may occur within a MG and to evaluate the feasibility of the control strategies to be derived.

因此,在这章中也充分地解决了微电源、储能设备和电力电子接口的建模,因为这是为了评估可能在微网中发生的动态相互作用和控制策略的可行性而必须的第一步。

[7] Europe, North America and Japan are leading the revolution being faced in the conventional electric power systems operation paradigm, by actively promoting research, development and deployment of MG.

通过积极推进微电网的研究、开发和部署,欧洲、北美和日本正在引领传统电力系统运行方式的改革。

[8] From the bulk power system perspective, the MG can establish contractual agreements for energy and possibly other services in a similar way to what happens with ordinary consumers or power producers.

从大容量电力系统的角度来看,微电网可以建立合同协议,这些协议是关于能源和以相同方式与普通消费者或生产者有关的其他服务。

[9] In case of large sites, it is suggested to divide the loads in many controllable units (buildings, industrial sites, etc.) and exploit the distribution system for the interconnection of several MG in order to supply the entire system.

在使用大型网络时,为了能供应整个系统,建议将负载分为许多可控的单位(如建筑物、工业用地等),使用多微电网互联的配电系统。

[10] The LV side of the distribution transformer is the Point of Common Coupling (PCC) between the MG and the distribution system and it is used to define the boundary between both systems.

在微电网与配电系统之间是配电变压器低压侧的公共耦合点(PCC),它用来定义两个系统之间的边界。

[11] The development of the protection system requires special attention due to the massive presence of power electronic converters, which have reduced capability to provide large fault currents.

由于电力电子变换器的大量存在,具有降低大的故障电流能力的保护系统的发展需要特别的关注。

[12] In order to achieve the desired flexibility, the MG system in centrally controlled and managed by the MicroGrid Central Controller (MGCC), installed in the LV side of the hierarchical level.

为了实现所需的灵活性,将微网中央控制器(MGCC)集中控制和管理的微电网系统安装在低压侧。

[13] The issues of autonomous and non-autonomous operation of the MG and the related exchange of information are examples of new important issues to be tackled in the near future.

微网的自主性和非自主运行的问题和相关的信息交换的例子是近期要解决的新的重要问题。

[14] A key functionality to be installed in the MGCC is forecasting of local loads and generation.

安装 MGCC 的一个重要作用是用来进行当地负荷和发电量的预测。

[15] In the Emergency Mode, an immediate change in the output power control of the MS is required, as they change from a dispatched power mode to one controlling frequency and voltage of the islanded section of the network.

在应急模式下，必须立即改变微电网的输出功率控制，因为它们改变了这段网络孤岛从一个发送功率到一个控制频率和电压的模式。

[16] When evaluating the quality of the communications channel, another important factor must also be considered: the level and nature of the interfering signals that are present at the input of the receiver.

评估通信信道的质量时，也必须考虑另一个重要因素：存在于接收机输入端的干扰信号的水平和性质。

Chapter 19   High Voltage Insulation

19.1   Introduction

Transient overvoltages caused by lightning strikes to transmission lines and by switching operations are of fundamental importance in selecting equipment insulation levels and surge-protection devices.[1]

The transmission line constants R, L, G, and C were recognized as distributed rather than lumped constants. When a line with distributed constants is subjected to a disturbance such as a lightning strike or a switching operation, voltage and current waves arise and travel along the line at a velocity near the speed of light.[2] When these waves arrive at the line terminals, reflected voltage and current waves arise and travel back down the line, superimposed on the initial waves.

Because of line losses, traveling waves are attenuated and essentially die out after a few reflections. Also, the series inductances of transformer windings effectively block the disturbances, thereby preventing them from entering generator windings. However, due to the reinforcing action of several reflected waves, it is possible for voltage to build up to a level that could cause transformer insulation or line insulation to arc over and suffer damage.[3]

Circuit breakers, which can operate within 50ms, are too slow to protect against lighting or switching surges. Lighting surges can rise to peak levels within a few microseconds and switching surges within a few hundred microseconds—fast enough to destroy insulation before a circuit breaker could open. However, protective devices are available. Called surge arresters, these can be used to protect equipment insulation against transient overvoltages. These devices limit voltage to a ceiling level and absorb the energy from lightning and switching surges.

19.2   Lightning

Cloud-to-ground (CG) lightning is the greatest single cause of overhead transmission and distribution line outages. Data obtained over a 14-year period from electric utility companies in the United States and Canada and covering 25,000 miles of transmission show that CG lightning accounted for about 26% of outages on 230kV circuits and about 65% of outages on 345kV circuits. A similar study in Britain, also over a 14-year period, covering 50,000 faults on distribution lines shows that CG lightning accounted for 47% of outages on circuits up to and including 33kV.

The electrical phenomena that occur within clouds leading to a lightning strike are complex and not totally understood. Several theories generally agree, however, that charge separation occurs within clouds. Wilson postulates that falling raindrops attract negative charges and therefore leave behind masses of positively charged air. The falling raindrops bring the negative charge to the bottom of the cloud, and upward air drafts carry the positively charged air and ice crystals to the top of the cloud, as shown in Fig. 19-1. Negative charges at the bottom of the cloud induce a positively charged region, or "shadow," on the earth directly below the cloud. The electric field lines shown in Fig. 19-1 originate from the positive charges and terminate at the negative charges.

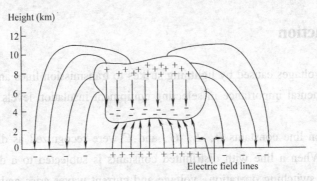

Fig. 19-1 Postulation of Charge Separation within Clouds

When voltage gradients reach the breakdown strength of the humid air within the cloud, typically 5kV/cm to 15kV/cm, an ionized path or downward leader moves from the cloud toward the earth. The leader progresses somewhat randomly along an irregular path, in steps. [4] These leader steps, about 50m long, move at a velocity of about 10^5m/s. [5] As a result of the opposite charge distribution under the cloud, another upward leader may rise to meet the downward leader. When the two leaders meet, a lightning discharge occurs, which neutralizes the charges.

The current involved in a CG lightning stroke typically rises to a peak value within 1 to 10μs, and then diminishes to one-half the peak within 20μs to 100μs. The distribution of peak currents is shown in Fig. 19-2. This curve represents the percentage of strokes that exceed a given peak current. For example, 50% of all strokes have a peak current greater than 45kA. In extreme cases, the peak current can exceed 200kA. Also, test results indicate that approximately 90% of all strokes are negative.

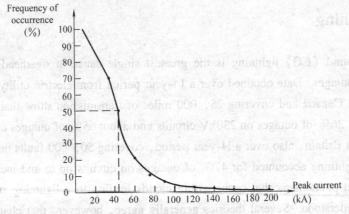

Fig. 19-2 Frequency of Occurrence of Lightning Currents That Exceed a Given Peak Value

It has also been shown that what appears to the eye as a single flash of lightning is often the cumulative effect of many strokes. A typical flash consists of typically 3 to 5, and occasionally as many as 40, strokes, at intervals of 50ms.

A typical transmission-line design goal is to have an average of less than 0.50 lightning outages per year per 100 miles of transmission. [6] For a given overhead line with a specified voltage rating, the following factors affect this design goal:

1) Tower height;
2) Number and location of shield wires;
3) Number of standard insulator discs per phase wire;
4) Tower impedance and tower-to-ground impedance.

It is well known that lightning strikes tall objects. Thus, shorter, H-frame structures are less susceptible to lightning strokes than taller, lattice towers. Also, shorter span lengths with more towers per kilometer can reduce the number of strikes.

Shield wires installed above phase conductors can effectively shield the phase conductors from direct lightning strokes. Fig. 19-3 illustrates the effect of shield wires. Experience has shown that the chance of a direct hit to phase conductors located within ±30° arcs beneath the shield wires is reduced. [7] Some lightning strokes are, therefore, expected to hit these overhead shield wires. When this occurs, traveling voltage and current waves propagate in both directions along the shield wire that is hit. When a wave arrives at a tower, a reflected wave returns toward the point where the lightning hit, and two refracted waves occur. One refracted wave moves along the shield wire into the next span. Since the shield wire is electrically connected to the tower, the other refracted wave moves down the tower, its energy being harmlessly diverted to ground. [8]

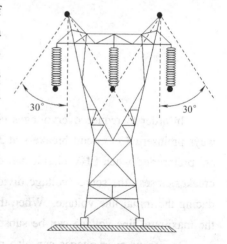

Fig. 19-3　Effect of Shield Wires

However, if the tower impedance or tower-to-ground impedance is too high, its voltages that are produced could exceed the breakdown strength of the insulator discs that hold the phase wires. The number of insulator discs per string is selected to avoid insulator flashover. Also, tower impedances and tower footing resistances are designed to be as low as possible. If the inherent tower construction does not give a naturally low resistance to ground, driven ground rods can be employed. Sometimes buried conductors running under the line (called *counterpoise*) are employed.

19.3　Switching Surges

The magnitudes of overvoltages due to lightning surges are not significantly affected by the power system voltage. On the other hand, overvoltages due to switching surges are directly proportional to system voltage. Consequently, lightning surges are less important for EHV transmission above 345kV and for UHV transmission, which has improved insulation. Switching surges become the limiting factor in insulation coordination for system voltages above 345kV.

One of the simplest and largest overvoltages can occur when an open-circuited line is energized, as shown in Fig. 19-4. Assume that the circuit breaker closes at the instant the sinusoidal source voltage has a peak value $\sqrt{2}$ V. Assuming zero source impedance, a forward traveling voltage wave of magnitude $\sqrt{2}$ V occurs. When this wave arrives at the open-circuited receiving end, where $\Gamma_R = +1$, the reflected voltage wave superimposed on the forward wave results in a maximum voltage of $2\sqrt{2}$ V = 2.83V. Even higher voltages can occur when a line is reclosed after momentary interruption.

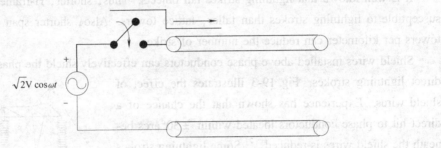

Fig. 19-4 Energizing an Open Circuited Line

In order to reduce overvoltages due to line energizing or reclosing, resistors are almost always preinserted in circuit breakers at 345 kV and above.[9] Resistors ranging from 200Ω to 800Ω are preinserted when EHV circuit breakers are closed, and subsequently bypassed. When a circuit breaker closes, the source voltage divides across the preinserted resistors and the line, thereby reducing the initial line voltage. When the resistors are shorted out, a new transient is initiated, but the maximum line voltage can be substantially reduced by careful design.

Dangerous overvoltages can also occur during a single line-to-ground fault on one phase of a transmission line. When such a fault occurs, a voltage equal and opposite to that on the faulted phase occurs at the instant of fault inception. Traveling waves are initiated on both the faulted phase and, due to capacitive coupling, the unfaulted phases.[10] At the line ends, reflections are produced and are superimposed on the normal operating voltages of the unfaulted phases.[11] Kimbark and Legate show that a line-to-ground fault can create an overvoltage on an unfaulted phase as high as 2.1 times the peak line-to-neutral voltage of the three-phase line.

19.4 Insulation Coordination

Insulation coordination is the process of correlating electric equipment insulation strength with protective device characteristics so that the equipment is protected against expected overvoltages.[12] The selection of equipment insulation strength and the protected voltage level provided by protective devices depends on engineering judgment and cost.

As shown by the top curve in Fig. 19-5, equipment insulation strength is a function of time. Equipment insulation can generally withstand high transient overvoltages only if they are of sufficiently short duration. However, determination of insulation strength is somewhat complicated. During repeated tests with identical voltage waveforms under identical conditions, equipment insulation may fail one test and withstand another.

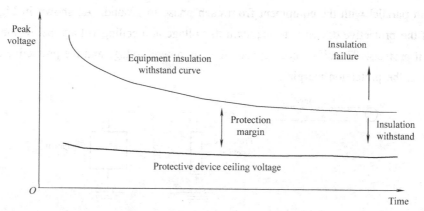

Fig. 19-5 Equipment Insulation Strength

For purposes of insulation testing, a standard impulse voltage wave, as shown in Fig. 19-6, is defined. The impulse wave shape is specified by giving the time T_1 in microseconds for the voltages to reach its peak value and the time T_2 for the voltage to decay to one-half its peak. One standard wave is a 1.2×50 wave, which rises to a peak value at $T_1 = 1.2 \mu s$ and decays to one-half its peak at $T_2 = 50 \mu s$.

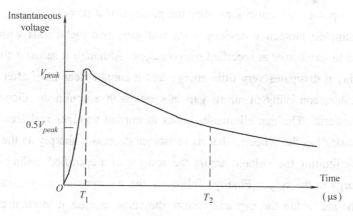

Fig. 19-6 Standard Impulse Voltage Waveform

Basic insulation level or BIL is defined as the peak value of the standard impulse voltage wave in Fig. 19-6. Equipment conforming to these BILs must be capable of withstanding repeated applications of the standard waveform of positive or negative polarity without insulation failure.[13] Also, these standard BILs apply to equipment regardless of how it is grounded.

Note that over-transmission-line insulation, which is external insulation, is usually self-restoring. When a transmission-line insulator string flashes over, a short circuit occurs. After circuit breakers open to deenergize the line, the insulation of the string usually recovers, and the line can be rapidly reenergized. However, transformer insulation, which is internal, is not self-restoring. When transformer insulation fails, the transformer must be removed for repair or replaced.

To protect equipment such as a transformer against overvoltages higher than its BIL, a protective device, such as that shown in Fig. 19-7, is employed. Such protective devices are generally connected in parallel with the equipment from each phase to ground. As shown in Fig. 19-5, the function of the protective device is to maintain its voltage at a ceiling voltage below the BIL of the equipment it protects. The difference between the equipment voltage and the protective device ceiling voltage is the protection margin.

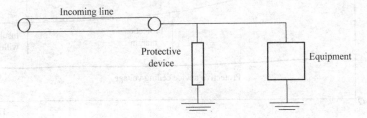

Fig. 19-7 Single-Line Diagram of Equipment and Protective Device

Protective devices should satisfy the following four criteria:

1) Provide a high or infinite impedance during normal system voltage, to minimize steady-state losses.
2) Provide a low impedance during surges, to limit voltage.
3) Dissipate or store the energy in the surge without damage to itself.
4) Return to open-circuit conditions after the passage of a surge.

One of the simplest protective devices is the rod gap, two metal rods with a fixed air gap, which is designed to spark over at specified overvoltages. Although it satisfies the first two protective device criteria, it dissipates very little energy and it cannot clear itself after arcing over.

A surge arrester, consisting of an air gap in series with a nonlinear silicon carbide resistor, satisfies all four criteria. The gap eliminates losses at normal voltages and arcs over during overvoltages. The resistor has the property that its resistance decreases sharply as the current through it increases, thereby limiting the voltage across the resistor to a specified ceiling. The resistor also dissipates the energy in the surge. Finally, following the passage of a surge, various forms of arc control quench the arc within the gap and restore the surge arrester to normal open-circuit conditions.[14]

The "gapless" surge arrester, consisting of a nonlinear metal oxide resistor with no air gap, also satisfies all four criteria. At normal voltage the resistance is extremely high, limiting steady-state currents to microamperes and steady-state losses to a few watts. During surges, the resistance sharply decreases, thereby limiting overvoltage while dissipating surge energy.[15] After the surge passes, the resistance naturally returns to its original high value. One advantage of the gapless arrester is that its ceiling voltage is closer to its normal operating voltage than the conventional arrester, thus permitting reduced BILs and potential savings in the capital cost of equipment insulation. [16]

There are four classes of surge arresters: station, intermediate, distribution, and secondary. Station arresters, which have the heaviest construction, are designed for the greatest range of ratings and have the best protective characteristics. Intermediate arresters, which have moderate construction, are designed for systems with nominal voltages 138kV and below. Distribution arresters are employed with lower-voltage transformers and lines, where there is a need for economy. Secondary arresters are used for nominal system voltages below 1000V.

New Words and Expressions

1. transient overvoltage   暂时过电压
2. lightning strike   雷击
3. switching operation   开关操作
4. surge-protection   过电压保护
5. distributed   *adj.* 分布式的
6. lumped   *adj.* 集中的
7. reflected wave   反射波
8. attenuate   *v.* 衰减
9. arc over   电弧放电，击穿
10. switching surge   操作过电压
11. surge arrester   避雷器
12. ceiling   *n.* 上限
13. charge separation   电荷分离
14. postulate   *v.* 假设
15. raindrop   *n.* 雨滴
16. air draft   气流
17. ice crystal   冰晶
18. electric field line   电场线
19. breakdown strength   击穿场强
20. ionize   *v.* 电离
21. neutralize   *v.* 中和
22. diminish   *v.* 减少
23. cumulative   *adj.* 累积的
24. shield wire   避雷线
25. insulator disc   碟形绝缘子
26. lattice tower   格构式塔架
27. span length   跨间距离
28. propagate   *v.* 传播
29. flashover   *n.* 闪络
30. EHV (extra high voltage) 超高压
31. UHV (ultra high voltage) 特高压
32. energized   *adj.* 通电的
33. reclose   *v.* 重合闸
34. traveling wave   行波
35. self-restoring   自恢复
36. deenergize   *v.* 断电
37. criteria   *n.* 标准
38. dissipate   *v.* 耗散，消散
39. silicon carbide   碳化硅
40. quench   *v.* 淬火

Notes

[1] Transient overvoltages caused by lightning strikes to transmission lines and by switching operations are of fundamental importance in selecting equipment insulation levels and surge-protection devices.

由雷击输电线和开关操作造成的暂态过电压，对于选择设备的绝缘水平和过电压保护装置非常重要。

[2] When a line with distributed constants is subjected to a disturbance such as a lightning strike or a switching operation, voltage and current waves arise and travel along the line at a veloc-

ity near the speed of light.

当一条具有分布常数的导线受到雷击或开关操作等干扰时，沿线产生和传播的电压和电流波的速度接近光速。

[3] However, due to the reinforcing action of several reflected waves, it is possible for voltage to build up to a level that could cause transformer insulation or line insulation to arc over and suffer damage.

然而，由于数次反射波的加强，有可能使电压增加到一定水平，从而导致变压器绝缘或线路绝缘产生电弧击穿，进而遭受损坏。

[4] When voltage gradients reach the breakdown strength of the humid air within the cloud, typically 5 to 15kV/cm, an ionized path or downward leader moves from the cloud toward the earth. The leader progresses somewhat randomly along an irregular path, in steps.

当云层的电压梯度达到潮湿空气的击穿场强（5~15kV/cm）时，形成从雷云到地面的电离路径或先导放电。先导只是无规则地向下逐步推进。

[5] These leader steps, about 50m long, move at a velcity of about 10^5m/s.

每级长度约50m，伸展速度约10^5m/s。

[6] A typical transmission-line design goal is to have an average of less than 0.50 lightning outages per year per 100 miles of transmission.

每年每100miles（1mile≈1.609km）输电线路的雷击停电次数平均不超过0.5次是一个典型传输线的设计目标。

[7] Experience has shown that the chance of a direct hit to phase conductors located within ±30° arcs beneath the shield wires is reduced.

经验表明，直接击中避雷线下方30°内的相线的概率会减少。

[8] Since the shield wire is electrically connected to the tower, the other refracted wave moves down the tower, its energy being harmlessly diverted to ground.

由于避雷线与铁塔相连，其他反射波的能量沿着塔被安全转移到地面。

[9] In order to reduce overvoltages due to line energizing or reclosing, resistors are almost always preinserted in circuit breakers at 345kV and above.

为了减少因线路通电或重合闸造成的过电压，通常在345kV及以上线路的断路器中预先插入合闸电阻。

[10] When such a fault occurs, a voltage equal and opposite to that on the faulted phase occurs at the instant of fault inception. Traveling waves are initiated on both the faulted phase and, due to capacitive coupling, the unfaulted phases.

单相接地故障时，在故障开始瞬间，会出现与故障相电压相等、方向相反的电压。行波开始在故障相传递，由于电容耦合非故障相中也同时传递。

[11] At the line ends, reflections are produced and are superimposed on the normal operating voltages of the unfaulted phases.

在线路两端，反射波产生并且将叠加于非故障相正常运行的电压上。

[12] Insulation coordination is the process of correlating electric equipment insulation strength with protective device characteristics so that the equipment is protected against expected

overvoltages.

绝缘配合就是协调电气设备的绝缘强度与保护装置的特性，从而保证设备能够抵御预期的过电压。

[13] Equipment conforming to these BILs must be capable of withstanding repeated applications of the standard waveform of positive or negative polarity without insulation failure.

符合这些基本绝缘水平的设备必须能够反复承受正负极性的标准波形，而不会出现绝缘故障。

[14] The resistor also dissipates the energy in the surge. Finally, following the passage of a surge, various forms of arc control quench the arc within the gap and restore the surge arrester to normal open-circuit conditions.

电阻器也能消耗浪涌中的能量。最后，随着浪涌的通过以及各种形式的电弧控制，使得电弧在间隙内熄灭并且将避雷器恢复到正常的开路条件。

[15] During surges, the resistance sharply decreases, thereby limiting overvoltage while dissipating surge energy.

在浪涌期间，电阻急剧下降，从而限制了过电压，同时消耗浪涌能量。

[16] One advantage of the gapless arrester is that its ceiling voltage is closer to its normal operating voltage than the conventional arrester, thus permitting reduced BILs and potential savings in the capital cost of equipment insulation.

无间隙避雷器相对于传统避雷器的一个优点是，它的上限电压接近其正常工作电压，从而允许降低基本绝缘水平和节约设备的绝缘费用。

Chapter 20 System Protection

20.1 Introduction

In this chapter we consider the problem of power system protection. Good design, maintenance, and proper operating procedures can reduce the probability of occurrence of faults, but cannot eliminate them. [1] Given that faults will inevitably occur, the objective of protective system design is to minimize their impact.

Faults are removed from a system by opening or "tripping" circuit breakers. These are the same circuit breakers used in normal system operation for connecting or disconnecting generators, lines, and loads. For emergency operation the breakers are tripped automatically when a fault condition is detected. Ideally the operation is highly selective; only those breakers closest to the fault operate to remove or "clear" the fault. The test of the system remains intact.

Fault conditions are detected by monitoring voltages and currents at various critical points in the system. Abnormal values individually or in combination cause relays to operate, energizing tripping circuits in the circuit breakers. A simple example is shown in Fig. 20-1.

A description of the operation is as follows. The primary current in the current transformer is the line current I_1. The secondary current I_1' is passed through the operating coil of an overcurrent relay. When $|I_1'|$ exceeds a specified pickup value, the normally open relay contacts close. If the relay is of the plunger type, the contacts close instantaneously. Other types of relays have an intentional and adjustable time delay. When the relay contacts close, the trip coil circuit is energized and the circuit breaker trips open. We can think of the trip coil as a solenoid whose movable core releases a latch, permitting the stored energy in a spring or compressed air tank to mechanically force open the movable contacts of the circuit breaker. We note in passing that the arc that forms between the circuit-breaker contacts when they open is extinguished by blowing away the ionized medium between the contacts using a blast of air or a transverse magnetic field; since the AC arc current is zero twice each cycle, the arc goes out once the insulating properties of the medium are restored. [2]

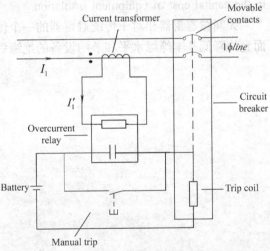

Fig. 20-1 Schematic of Overcurrent Protection

20.2 Protection of Radial Systems

We start by considering the simplest case, that of protecting a radial system, as shown in Fig. 20-2. The protection scheme specified should satisfy some general criteria assuring selectivity and redundancy. More specifically in this particular case, we wish to satisfy the following criteria:

1) Under normal load conditions the circuit breakers B_0, B_1 and B_2 do not operate.

2) Under fault conditions only the closest breaker to the left of the fault should operate; this protects the system by clearing (and reenergizing) the fault while maintaining service to as much of the system as possible.

3) In the event that the closest breaker fails to operate, the next breaker closer to the power source should operate. [3] This provision is called backup protection.

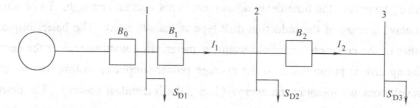

Fig. 20-2 Radial System

Suppose that there is a 3-phase fault to the right of B_2. The current I_2, at B_2, will increase from its prefault value to a, presumably, much larger fault value. The presence of this large current can be detected by an overcurrent relay at the breaker location and used to initiate the triggering of B_2. Stated more simply and directly in breaker terms, we can trip B_2 when its fault current magnitude exceeds a specified preset pickup value of current. If we set the pickup value of current low enough to trip B_2 for a fault at bus 3, then B_2 will certainly trip for the larger current accompanying a "closer in" fault anywhere on the transmission line to the right of B_2. Thus B_2 protects the line and bus to its right, or stated differently, the transmission line and bus are in the protection zone of B_2.

What if B_2 fails to operate for a fault in its protection zone? We note that the fault current at B_1 is essentially the same as the fault current at B_2, since the load current (supplying S_{D2}) is small compared with the fault current. We could therefore set B_1 for the same pickup current value as B_2 but introduce a time delay so that B_2 would normally operate first; this deliberately introduced time delay is called a coordination time delay. Would this method of setting the pickup current value of B_1 interfere with its primary function in protecting against faults in its protection zone (line and bus between B_1 and B_2)? Fortunately not, since faults within its protection zone are closer to the source and produce larger fault currents. If B_1 is set to trip for remote faults (in the protection zone of B_2), it certainly can be expected to trip for faults within its own protection zone. The preceding discussion for the two breakers B_1 and B_2 extends pairwise to any number of breakers in a radial system. A difficulty in extending the number of breakers, however, is the possibly excessive trip time for the breakers nearest the source because of the summation of coordination time delays. [4]

While for simplicity we have been discussing a symmetric 3-phase fault, we certainly will want to protect our system against the more common general faults. The extension is simple enough. For example, for B_2, the pickup current for each phase should be low enough that tripping occurs for any type fault of interest in its zone of protection. [5] The implementation will require three overcurrent relays with their contacts in parallel in the trip circuit of the 3-phase circuit breaker. Thus, a fault on any phase trips the breaker. This practice of tripping all three phases is common in the United States. In some other countries single-pole switching may be used for some types of faults.

We note that the plunger-type relay can be used as an overcurrent relay with adjustable pickup current. It can also be fitted with an oil dash pot or air bellows to provide an adjustable time delay. In practice, however, the time-delay adjustment is not accurate enough. To obtain a more accurate time delay, a relay of the induction-disk type is usually used. The basic torque generating principle is that of the common household watthour meter. In a watthour meter the angular velocity of the rotating disk is proportional to the average power supplied; in this way by counting the number of revolutions we measure the energy (i.e., the integral of power). The desired characteristic is achieved by designing the watthour meter so that the torque that drives the disk is proportional to the average power supplied:

$$T = k|V||I|\cos\phi \qquad (20-1)$$

where ϕ is the phase angle between V and I. There are also damping or braking magnets across the disk that cause an opposing torque proportional to the speed. Thus, in equilibrium we get the desired proportionality between angular velocity and average power.

Although the actual construction of an induction-disk relay differs in detail, we can think of using a modified watthour meter element with the voltage and current coils connected externally to give a torque proportional to $|I|^2$ alone. [6] If the disk is then restrained by a spiral spring, it will not turn until the current exceeds a certain pickup value. Beyond that value the higher the current magnitude, the faster the disk rotates and the less time it takes to rotate through a given angle. By having a moving contact that rotates with the disk "makeup" with a stationary contact, an "inverse" time-current characteristic may be obtained. [7] By adjusting the position of the stationary contact, the time delay may be varied and a family of time-current characteristics is obtained.

To obtain an adjustable pickup current value, one might make the spiral spring tension adjustable. In fact, the adjustment is obtained by varying the number of turns in the current coil of the relay by means of "taps".

20.3 System with Two Sources

Let us consider what happens if we modify the system in Fig. 20-2 by adding a source to Bus 3. The advantage of the modified system compared with a radial system is improved continuity of service because the power can be supplied from either end. Additional circuit breakers are needed as shown in Fig. 20-3.

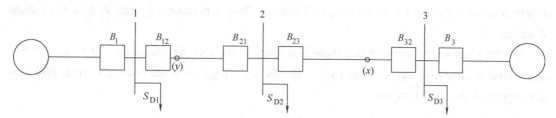

Fig. 20-3 System with Two Sources

Note that a fault on Bus 1, or the line between Bus 1 and Bus 2, need no longer interrupt service to Bus 2 and Bus 3. This, of course, presumes correct operation by the breakers.

Correct operation cannot be accomplished by the method discussed in the preceding section. To see this, suppose that there is a solid 3-phase fault at point (x) in Fig. 20-3. We want B_{23} and B_{32} to open to clear the fault. We do not want B_{12} and /or B_{21} to operate and interrupt service to Bus 2. Suppose that we attempt to use the method of the preceding section. We would set B_{23} to trip faster than B_{21} (and B_{21} to trip faster than B_{12}), which would imply correct action for a fault at point (x). But suppose that the fault were at point (y) instead. With relays set as described previously, B_{23} would trip before B_{21}, which would isolate Bus 2 unnecessarily. When the fault may be fed from either left or right, there is no way to coordinate the relay time delays properly.

To remove faulted lines correctly, we require relays to respond only to faults occurring on their forward or line sides. [8] The relays must therefore be directional. In this case B_{21} would not operate for a fault at location (x); B_{23} and B_{12} could be coordinated so that B_{23} would operate first. On the other hand, with a fault at location (y), B_{23} would not operate and B_{21} and B_{32} could be coordinated so that B_{21} would operate first. In fact, with this directional feature, and for a system like that in Fig. 20-3 (i. e., with a string of lines and buses fed from both ends), the coordination may be successfully accomplished just as in the case of a radial system, and the two breakers closest to the fault would be expected to operate first. Note that these two breakers would not generally operate simultaneously since the short-circuit currents and coordination time delays would be different for the breakers to the left and the right of the fault point.

Note, finally, that with directional relays the buses themselves are still protected. A fault at Bus 2, for example, is on the line sides of B_{12} and B_{32} and would be "seen" by these breakers.

We next consider how this directional feature may be obtained. It is easiest to explain the basic idea by considering only solid 3-phase faults. Suppose that in Fig. 20-3 at B_{23}, using instrument transformers, we can measure the following phase a quantities: V_2, the phase-neutral voltage, and I_{23}, the line current flowing from Bus 2 to Bus 3.

Suppose that there is a symmetric 3-phase fault on the line to the right of B_{23} [i. e., in the forward direction at some point such as (x)]. Then I_{23} will be in the form

$$I_{23} = \frac{V_2}{\lambda Z} \qquad (20\text{-}2)$$

where Z is the total series impedance of the line between Buses 2 and 3 and λ is the fraction of the total line between the breaker and the fault (i.e., a number between 0 and 1). $Z = |Z| \angle Z$ is mostly reactive with $\angle Z = \theta_z$ in the range 80 to 88°. Thus in practice, I_{23} lags V_2 by a large angle of almost 90°.

On the other hand, suppose that there is a solid 3ϕ fault in the line to the left of B_{23} [i.e., in the reverse direction, say at point (y)]. In this case, neglecting the relatively small load current supplying S_{D2}, we find that

$$I_{23} = -I_{21} = -\frac{V_2}{\lambda Z'} \qquad (20\text{-}3)$$

where Z' is the total series impedance of the line between Buses 1 and 2, and λ has the same meaning as before. $\angle Z'$ will be in the same range as $\angle Z$ (i.e., in the range 80° to 88°); thus from Equation (20-3) we see that in this case I_{23} leads V_2 by an angle between 92° and 100°. Thus the phase of I_{23} relative to that of V_2 is a distinctive feature that can be used to determine the fault direction.

20.4 Impedance (Distance) Relays

There are some problems with coordinating time-delay over current relays even in the case of radial systems. If the string of lines and buses is too long, the time for operation of breakers close to the source becomes too large. To overcome this problem a different principle can be employed. It is true that over current is a characteristic feature indicating a fault condition. An even more distinctive feature, however, is the ratio of voltage to current magnitudes. When there is a symmetric 3-phase fault, the voltage drops and the current rises. Suppose that the voltage drops to 50% of its normal value while the current increases to 200% of its normal value. Then there is a 4 : 1 change in the voltage-to-current ratio compared with only a 2 : 1 change in the current.

A relay that operates on the basis of a voltage-to-current ratio is naturally called a ratio or impedance relay. It is also called a distance relay; the reason will be seen when we describe its operation.

Consider Fig. 20-3 again and suppose that there is a solid 3-phase fault at point (x). Assume again that at B_{23} we measure the phase-neutral voltage V_2 and the current I_{23}. In this case the ratio V_2/I_{23} is the driving-point impedance of the portion of the line between B_{23} and the fault point (x):

$$\frac{V_2}{I_{23}} = \lambda Z \qquad (20\text{-}4)$$

where Z is the total line impedance and λ is the fraction of the line between B_{23} and the fault. In general, V_2/I_{23} is not a driving-point impedance and it is probably less confusing to consider it simply as a measured quantity at the observation point.

The relay is designed to operate if $|V_2/I_{23}|$ is small enough. Suppose that we wish 80% of the line to be in the zone of protection; then the relay would be designed to operate if $|V_2/I_{23}| \leq$

$R_C \triangleq 0.8|Z|$. The relay would then operate for any (solid 3-phase) fault located within the nearest 80% of the line (i.e., within the distance 80% of the length of the line). This explains the designation "distance" relay. We also speak of the reach of the relay; in this case the reach of the relay is 80% of the length of the line.

The region of operation of the relay is seen to be any complex value of V_2/I_{23} that lies within a radius R_C of the origin of the complex plane. In Fig. 20-4 we show several possible outcomes (i.e., values of V_2/I_{23} under different conditions); the conditions refer to the fault points labeled in Fig. 20-5. These points may be described as follows:

(a) Fault on line at 60% point (trip)

(b) Fault on line at 80% point (marginal trip)

(c) Fault on line at 100% point (block)

(d) Fault on system just beyond protected line (block)

(e) Fault on system much beyond protected line (block)

(f) Typical normal operating condition (block)

(g) Fault on line to the left of protected line (trip)

(h) More distant fault in same line as (g) (block)

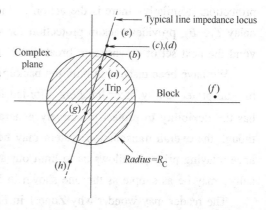

Fig. 20-4 Zone of Operation of Impedance Relay

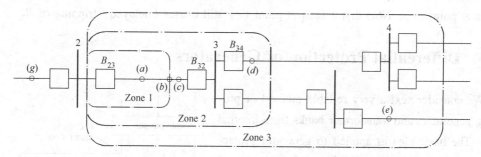

Fig. 20-5 Fault Points and Protected Zones

To avoid an undesired trip under condition (g) we can include a directional relay, just as we did previously in the case of over-current relays. Also, as with the over-current relays, we can provide backup protection. Consider a section of the transmission system as shown in Fig. 20-5. We are considering only the relaying at the B_{23} location for the purpose of tripping B_{23}. We note that the other breakers (for example, B_{34}) are similarly equipped.

Assuming the use of directional relay, we need only consider the system to the right of B_{23}. At B_{23} we install three impedance relays. The first, already described, has a reach out to, say, 80% of the line and is set to trip instantaneously if $|V_2/I_{23}| \leq R_1$. This protected region is designated Zone 1; it is also described as the primary, or instantaneous trip, Zone. The second impedance relay has a longer reach; it includes all of Zone 2 (which contains Zone 1). If there is a fault at point (d), for example, then after a time delay T_2 the breaker B_{23} will open. Presumably, point (d) is in the primary protection zone of breaker B_{34} and therefore B_{34} should open instantaneously before B_{23} has opened. However, in the event B_{34} does not open, B_{23} will provide backup protection. Similarly, there is the action of the third impedance relay. After an even longer time delay T_3, B_{23} provides backup protection for a Zone 3 fault (for example, at point (e) just beyond the next set of downstream breakers).

We have been discussing only the backup protection afforded by B_{23}. There is, of course, the backup protection by other breakers afforded to B_{23}. We note that directional-impedance relaying has the flexibility to protect effectively general transmission systems, not just radial ones.[9] Although the overall transmission system may be very complex, including many loops, the impedance relaying principle allows us to limit our attention to a much smaller domain. The system, locally, may be as simple as the one shown in Fig. 20-5.

The reader may wonder why Zone 1 in Fig. 20-5 was not specified to include the entire transmission line (i.e., 100% of the line). The reason may be explained as follows. The impedance seen by the impedance relay for a fault at a point (c) just to the left of Bus 3 and at a point (d) just to the right of the bus is essentially the same; the impedance of the closed breakers and the bus structure is negligible.[10] Thus if point (c) were included in Zone 1 (instantaneous trip), a fault at point (d) would also cause the instantaneous trip of B_{23}. This would be an undesirable trip since B_{34} should be allowed to clear the fault. The advantage of not tripping B_{23} when the fault is at point (d) more than compensates for the disadvantage of not tripping instantaneously when the fault is at point (c). Note that a fault at point (c) still causes (delayed) tripping of B_{23}.

20.5 Differential Protection of Generators

We consider next a very reliable method of protecting generators and transformer banks from internal faults. The basic idea as applied to generator protection may be seen in Fig. 20-6. Only the protection of phase a is shown. The scheme is repeated for the other phases. Using the dot convention, the reader may check that the reference directions for primary and secondary current in the CTs is consistent.

If there is no internal fault, $I_2 = I_1$, and with identical current transformers, $I_2' = I_1'$. Since the CTs are connected in series, the current flows one secondary to the other, and there is no current in the operating winding of the relay. Now suppose that there is a ground in the winding; or a phase-phase short. Then

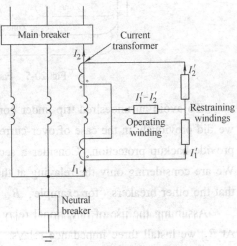

Fig. 20-6 Generator Protection by Differential Relaying

$I_2 \neq I_1$, and thus $I'_2 \neq I'_1$ and a differential current $I'_1 - I'_2$ flows in the operating winding. It is natural to call such a relay a differential relay.

It would seem reasonable to detect the differential current with an over-current relay, but in practice this is not done. The reason is the difficulty in matching the current transformer characteristics; they obviously cannot be identical. Even very small mismatches in, say, the turns ratio might cause significant differential current under full-load conditions or in the case of high currents due to fault conditions elsewhere. [11] On the other hand, raising the pickup value for the over-current relay to prevent improper operation at high generator currents would degrade the protection of the generator under light-load conditions. [12]

A solution to the problem is achieved by using a proportional or percentage type of differential relay, where the differential current required for tripping is proportional to the generator current. A balanced-beam movement with a center-tapped restraining winding can be used to obtain this behavior. The general idea is shown in Fig. 20-7.

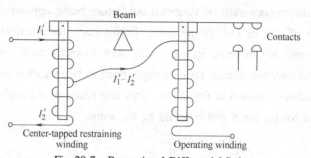

Fig. 20-7 Proportional Differential Relay

As in previous cases, the relay closes if the downward pull on the right side exceeds the downward pull on the left. The average downward pull on the right is proportional to the square of the differential current (i.e., to $|I'_1 - I'_2|^2$). The average downward pull on the left (the restraining force) is proportional to $|I'_1 + I'_2|^2$. Thus the condition for operation can be stated as

$$|I'_1 - I'_2| \geq k \frac{|I'_1 + I'_2|}{2} = k|I'_{\text{average}}| \tag{20-5}$$

and has the desired proportional property. k is a constant that is adjustable by tap and air-gap (core screw) adjustments. Notice that as k is increased the relay becomes less sensitive.

Assuming that I'_1 and I'_2 are in phase, it is easy to find the region of operation given by Equation (20-5). This is shown in Fig. 20-8 for a value of $k = 0.1$. We note that the characteristic can be easily modified to give an unambiguous block zone in the neighborhood of the origin. We also note that as an alternative to the balanced-beam relay we have been describing, there are induction-type relays that also operate on differential currents and have restraining windings. They have similar characteristics to Fig. 20-8.

We have been considering the differential protection of

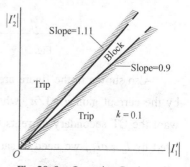

Fig. 20-8 Operating Region of Differential Relay

phase a of the generator winding; phases b and c are similarly protected. When any one of the three relays operates, the 3-phase line (main) circuit breaker and the neutral breaker open, isolating the generator from the rest of the system. The breakers, but not the tripping circuits, are shown in Fig. 20-6. In addition, the generator field breaker (not shown) would be tripped to deenergize the stator windings. This protects the generator in the event of (internal) phase-phase short, which would not be isolated with the opening of the main and neutral breakers. Since we need to open three different circuit breakers (main, neutral, and field), three sets of contacts must be available on each differential relay.

20.6 Differential Protection of Transformers

To protect a 3-phase transformer bank is somewhat more complicated because we must take into account the change in current magnitude and phase in going from the low-voltage to the high-voltage side.[13] The differences must be canceled out before being applied to a differential relay. Consider first the case of Y-Y or D-D connections. In this case, for practical purposes, the primary and secondary currents are in phase in the steady state (exactly so if the magnetizing inductances are infinite) and only the current ratio changes need to be canceled out. A possible realization for a Y-Y connection is shown in Fig. 20-9. Only one phase of the 3-phase relay protection is shown; the protection for phases b and c would be the same.

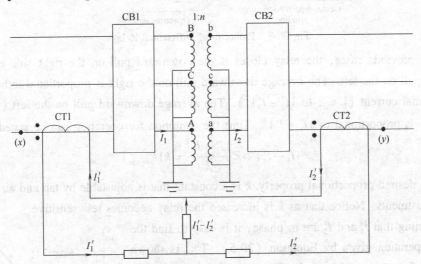

Fig. 20-9 Differential Protection of Y-Y Transformer Bank

Also shown in the figure are the phase a transformer bank currents I_1 and I_2, which are related by the current gain $a = 1/n$, where n is the voltage gain. It under normal, non-trip conditions we want the CT secondary currents to be equal (i.e., $I_2' = I_1'$) while the primary CT currents are related by $I_2 = aI_1$, we need to satisfy the following condition:

$$\frac{I_2'}{I_2} = \frac{I_1'}{aI_1} \tag{20-6}$$

Since $a_2 \triangleq I_2'/I_2$ and $a_1 \triangleq I_1'/I_1$ are the current gains of CT2 and CT1, respectively, we get the following simple relation between current gains:

$$a = \frac{a_1}{a_2} \qquad (20\text{-}7)$$

From Equation (20-7) the desired CT ratings may be selected. For example, suppose that the voltage gain or step up ratio $n = 10$. Then $a = a_1/a_2 = 0.1$. We might pick the primary rating of CT1 to be 500; thus $a_1 = 5/500 = 0.01$. Picking the primary rating of CT2 to be 50, we get $a_2 = 5/50 = 0.1$. Thus $a_1/a_2 = 0.1$, as required. The selection of CT ratings is natural; we need a higher rating (10 times higher) in the high-current side of the 3-phase transformer bank than in the low-current side.

In other cases there may be difficulty in finding standard CTs with the necessary ratios. Other methods for accomplishing the same purpose include the use of autotransformers connected to the CTs with multiple taps that provide additional flexibility.[14] Taps may also be provided in the differential relay itself.

There is an interpretation of the condition Equation (20-7) implying zero differential current, which will be helpful in considering the next case: the differential protection of D-Y transformer banks. Consider the current gain in going from point (x) to point (y) in Fig. 20-9. There are two possible paths. If we go through the 3-phase transformer bank, the gain is $I_2/I_1 = a$. If we go through the CTs, the gain is $(I_1'/I_1)(I_2/I_2') = a_1/a_2$. Thus to get zero differential current, Equation (20-7) says that the current gain must be the same for the two parallel paths. It is easy to show this condition must be true even if the current gains are complex numbers. The reader may note the connection with "normal" systems. Equation (20-7) says that the CT ratios and phase shifts should be in accordance with that of a normal system.

In Fig. 20-10 we show the differential protection of a D-Y transformer bank. The basic idea can best be understood for a one-line diagram.

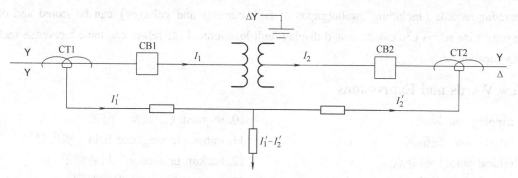

Fig. 20-10 Differential Protection of D-Y Transformer Bank

The point to note is that the secondaries of CT2 are connected in D. Thus when we consider the phase a (per phase) current gain a_2, we get a phase shift just as we do in the case of the D-Y power transformer bank; Equation (20-7) requires that the phase shifts be the same from the D to the Y sides. By wiring the sets of D-Y transformers identically, we get the same (connection-induced) phase shifts in going from the D to the Y sides. Of course, the current-gain magnitudes must also be equal, as required by Equation (20-7).

20.7 Computer Relaying

The electromechanical relays we have been discussing are rapidly being replaced by microprocessor-based relays. This is particularly the case in installations. For the most part the new relays replicate the functions and operating characteristics of the old relays.[15] Thus the protection techniques we have discussed involving backup protection and relay (time) coordination, etc., carry over unchanged. The physical implementation, however, is very much changed.

Some of the advantages of microprocessor-based relays include the following:

1) Low cost.

2) Low inventory. A few standard products may be variously and easily configured.

3) Low maintenance.

4) Very flexible tripping characteristics that can be changed with software and form a remote location.

5) Self-monitoring with the ability to report relay failure automatically.

6) A single relay with a single set of inputs can be programmed to carry out a number of functions (e.g., distance protection for multiple zones). Instead of, say, three impedance relays, each with a different reach, to cover three different zones; we can combine the three functions into a single unit.

7) Reduced panel space and station power.

8) Additional functions. In addition to their primary function of providing protection, microprocessor relay packages can also provide communications and recordkeeping.[16] Voltage and current readings and relay and apparatus status can be transmitted to system operators. Detailed fault operating records (including oscillographs of fault currents and voltages) can be stored and displayed. Line relays can calculate and display fault locations. Line relays can have a revenue metering capability.

New Words and Expressions

1. tripping   n. 跳闸
2. intact   adj. 完整的
3. critical point   临界点
4. relay   n. 继电器
5. operating coil   工作线圈
6. relay contact   继电器触点
7. plunger type   柱塞型
8. solenoid   n. 螺旋管
9. latch   n. 门闩，闩

10. in passing   顺便，附带
11. transverse magnetic field   横向磁场
12. backup protection   后备保护
13. protection zone   保护范围
14. pairwise   adj. 两两的
15. dash pot   减震器，缓冲器
16. damping   n. 阻尼
17. braking magnet   制动磁铁
18. opposing torque   反向转矩

19. spiral spring　　螺旋形弹簧
20. tension　　*n.* 张力
21. directional relay　　方向继电器
22. impedance relay　　阻抗继电器
23. driving-point　　驱动点
24. designation　　*n.* 名称
25. dot convention　　同名端协定
26. differential protection　　差动保护
27. phase-phase short　　相间短路
28. restraining winding　　制动线圈
29. cancel out　　使平衡（抵消）
30. autotransformer　　*n.* 自耦变压器
31. microprocessor-based relay　　微机继电器
32. recordkeeping　　记录保存

Notes

[1] Good design, maintenance, and proper operating procedures can reduce the probability of occurrence of faults, but cannot eliminate them.

好的设计、维护和适当的操作程序可以减少故障发生的概率，但不能完全避免故障的发生。

[2] We note in passing that the arc that forms between the circuit-breaker contacts when they open is extinguished by blowing away the ionized medium between the contacts using a blast of air or a transverse magnetic field; since the AC arc current is zero twice each cycle, the arc goes out once the insulating properties of the medium are restored.

在断路器触点打开时形成的电弧，是通过用空气流或者横向磁场吹走触点间的电离介质来实现灭弧的。因为在一个周期内，交流电弧电流两次达到零值，所以当设备的绝缘性能恢复时，电弧就会熄灭。

[3] In the event that the closest breaker fails to operate, the next breaker closer to the power source should operate.

在最近的断路器未能动作的情况下，接近电源的下一个断路器应该实施动作。

[4] A difficulty in extending the number of breakers, however, is the possibly excessive trip time for the breakers nearest the source because of the summation of coordination time delays.

但是，在增加断路器数量方面存在困难，因为这些断路器延迟时间的总和可能会超过离电源最近的断路器的跳闸时间。

[5] For example, for B_2, the pickup current for each phase should be low enough that tripping occurs for any type fault of interest in its zone of protection.

以 B_2 为例，每相的启动电流应该足够低，才能使它保护区域内发生任何类型的故障都会跳闸。

[6] Although the actual construction of an induction-disk relay differs in detail, we can think of using a modified watthour meter element with the voltage and current coils connected externally to give a torque proportional to $|I|^2$ alone.

尽管感应圆盘继电器的实际结构不尽相同，我们可以考虑使用外接有电压和电流线圈元件的改进电表，来提供只与 $|I|^2$ 成比例的转矩。

[7] By having a moving contact that rotates with the disk "makeup" with a stationary contact, an "inverse" time-current characteristic may be obtained.

通过使用活动触点旋转圆盘来配合静触点，就可以获得一个反时限电流特性。

[8] To remove faulted lines correctly, we require relays to respond only to faults occurring on their forward or line sides.

为了能够正确切除故障线路，我们需要继电器只对它们前端或两侧的线路故障产生响应。

[9] We note that directional-impedance relaying has the flexibility to protect effectively general transmission systems, not just radial ones.

我们可以得出，方向阻抗继电器灵活性能够有效地保护一般的传输系统，而不仅仅是单方向的。

[10] The impedance seen by the impedance relay for a fault at a point (c) just to the left of Bus 3 and at a point (d) just to the right of the bus is essentially the same; the impedance of the closed breakers and the bus structure is negligible.

故障位于母线 3 的左侧 c 点和右侧 d 点时，由阻抗继电器测得的等效阻抗基本相同。闭合的断路器和母线的阻抗可以忽略。

[11] Even very small mismatches in, say, the turns ratio might cause significant differential current under full-load conditions or in the case of high currents due to fault conditions elsewhere.

也就是说，即使存在非常小的不匹配，在满负荷条件下或由于故障造成高电流时，匝数比都可能造成非常大的差动电流。

[12] On the other hand, raising the pickup value for the overcurrent relay to prevent improper operation at high generator currents would degrade the protection of the generator under light-load conditions.

另一方面，为了防止在高电流下的误动作，而提高过电流继电器的动作值，会降低在轻载条件下发电机的保护能力。

[13] To protect a 3-phase transformer bank is somewhat more complicated because we must take into account the change in current magnitude and phase in going from the low-voltage to the high-voltage side.

三相变压器组的保护必须考虑从低压到高压侧的电流幅值和相位的变化，因此会更复杂一些。

[14] Other methods for accomplishing the same purpose include the use of autotransformers connected to the CTs with multiple taps that provide additional flexibility.

其他可以实现相同目的的方法包括将自耦变压器连接到有多个分接头的电流互感器，这样可以提供更多的灵活性。

[15] For the most part the new relays replicate the functions and operating characteristics of the old relays.

在大多数情况下，新的继电器会采用旧继电器的功能和工作特性。

[16] Additional functions. In addition to their primary function of providing protection, microprocessor relay packages can also provide communications and recordkeeping.

附加的功能。微机继电保护装置除了主要提供保护功能以外，还可以通信和保存记录。